Solid State Chemistry

Solid State Chemistry

An introduction

Lesley Smart and Elaine Moore
*Department of Chemistry,
The Open University, UK*

CHAPMAN & HALL
University and Professional Division

London · Glasgow · New York · Tokyo · Melbourne · Madras

Published by Chapman & Hall, 2–6 Boundary Row, London SE1 8HN

Chapman & Hall, 2–6 Boundary Row, London SE1 8HN, UK

Blackie Academic & Professional, Wester Cleddens Road, Bishopbriggs, Glasgow G64 2NZ, UK

Chapman & Hall, 29 West 35th Street, New York NY10001, USA

Chapman & Hall Japan, Thomson Publishing Japan, Hirakawacho Nemoto Building, 6F, 1-7-11 Hirakawa-cho, Chiyoda-ku, Tokyo 102, Japan

Chapman & Hall Australia, Thomas Nelson Australia, 102 Dodds Street, South Melbourne, Victoria 3205, Australia

Chapman & Hall India, R. Seshadri, 32 Second Main Road, CIT East, Madras 600 035, India

First edition 1992
© 1992 Lesley Smart and Elaine Moore

Typeset in 10/12pt Times by Excel Typesetters Ltd
Printed in Singapore by Fong & Sons Printers Pte Ltd

ISBN 0 412 40040 5

A catalogue record for this book is available from the British Library

Library of Congress Cataloging-in-Publication data
Smart, Lesley.
    Solid state chemistry: an introduction / Lesley Smart and Elaine Moore.
        p.   cm.
    Includes bibliographical references (p.   ) and index.
    ISBN 0–412–40040–5 (pb.)
    1. Solid state chemistry.   I. Moore, Elaine (Elaine A.)   II. Title.
QD478.S53   1992
541′.0421—dc20   91-41164
                                                                          CIP

Dedicated
to
Graham, Sam, Rosemary and Laura

Contents

Preface

The idea for this book originated with our involvement in an Open University inorganic chemistry course (S343: Inorganic Chemistry). When the Course Team met to decide the contents of this course, we felt that solid state chemistry had become an interesting and important area that must be included. It was also apparent that this area was playing a larger role in the undergraduate syllabus at many universities, due to the exciting new developments in the field.

Despite the growing importance of solid state chemistry, however, we found that there were few textbooks that tackled solid state theory from a chemist's rather than a physicist's viewpoint. Of those that did, most if not all, were aimed at final year undergraduates and postgraduates. We felt there was a need for a book written from a chemist's viewpoint that was accessible to undergraduates earlier in their degree programme. This book is an attempt to provide such a text.

Because a book of this size could not cover all topics in solid state chemistry, we have chosen to concentrate on structures and bonding in solids, and on the interplay between crystal and electronic structure in determining their properties. Examples of solid state devices are used throughout the book to show how the choice of a particular solid for a particular device is determined by the properties of that solid.

Chapter 1 is an introduction to crystal structures and the ionic model. It introduces many of the crystal structures that appear in later chapters and discusses the concepts of ionic radii and lattice energies. Ideas such as close-packed structures and tetrahedral and octahedral holes are covered here; these are used later to explain a number of solid state properties.

Chapter 2 introduces the band theory of solids. The main approach is via the tight binding model, seen as an extension of the molecular orbital theory familiar to chemists. Physicists more often develop the band model via the free electron theory, which is included here for completeness. This chapter also discusses electronic conductivity in solids and in particular properties and applications of semiconductors.

Chapter 3 is concerned with solids that are not perfect. The types of defect that occur and the way they are organized in solids forms the main subject matter. Defects lead to interesting and exploitable properties and several examples of this appear in this chapter, including photography and solid state batteries.

The remaining chapters each deal with a property or a special class of solid. Chapter 4 covers low-dimensional solids, the properties of which are not isotropic. Chapter 5 deals with zeolites, an interesting class of compounds used extensively in industry (as catalysts for example), the properties of which strongly reflect their structure. Chapter 6 deals with optical properties and Chapter 7 with magnetic properties of solids. Finally, Chapter 8 explores the exciting field of superconductors, in particular the relatively recently discovered high-temperature superconductors.

The approach adopted is deliberately non-mathematical, and assumes only the chemical ideas that a first year undergraduate would have. For example, differential calculus is used on only one or two pages and non-familiarity with this would not hamper an understanding of the rest of the book; topics such as ligand field theory are not assumed.

As this book originated with an Open University text, it is only right that we should acknowledge the help and support of our colleagues on the Course Team, in particular Dr David Johnson and Dr Kiki Warr. We are also grateful to Dr Joan Mason who read and commented on much of the script and to the anonymous reviewer to whom Chapman and Hall sent the original manuscript and who provided very thorough and useful comments.

The authors have been sustained through the inevitable drudgery of writing, by an enthusiasm for this fascinating subject. We hope that some of this transmits itself to the student.

Lesley E. Smart and Elaine A. Moore
OU, Walton Hall, Milton Keynes

An introduction to simple crystal structures

1.1 INTRODUCTION

All substances if cooled sufficiently form a solid phase; the vast majority form one or more **crystalline** phases, where the atoms, molecules or ions pack together to form a regular repeating array. This book is concerned mostly with the structure of metals and of ionic solids; these do not contain discrete molecules as such, but comprise extended arrays of atoms or ions.

Crystal structures are usually determined by the technique of **X-ray crystallography**. This technique relies on the fact that the atomic spacings in crystals are of the same order of magnitude as the wavelength of X-rays (of the order of $1\,\text{Å}$ or $100\,\text{pm}$): a crystal thus acts as a three-dimensional diffraction grating to a beam of X-rays. The resulting diffraction pattern can be interpreted to give the internal positions of the atoms in the crystal very precisely. We will not attempt to describe in detail this, or any other, physical technique in this book, but only to use the results. The further reading lists at the end of each chapter give references to suitable texts or articles on physical techniques.

The structures of many inorganic solids can be discussed in terms of the simple packing of spheres, so it is at this point we will start, before moving on to the more formal classification of crystals.

1.2 CLOSE PACKING

Think for the moment of an atom as a small hard sphere. Figure 1.1 shows two possible arrangements for a layer of such identical atoms. On squeezing the square layer in Figure 1.1a, the spheres would move to the positions in Figure 1.1b so that the layer takes up less space. The layer in Figure 1.1b (layer A) is called **close-packed**.

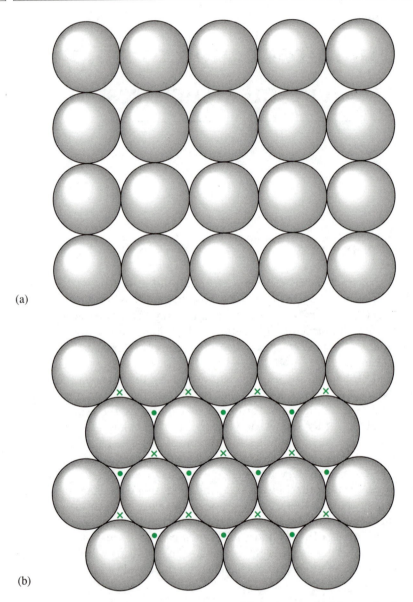

(a)

(b)

Figure 1.1 (a) A square array of spheres. (b) A close-packed layer of spheres.

To build up a close-packed structure in three dimensions we must now add a second layer (layer B). The spheres of the second layer sit in half of the hollows of the first layer: these have been marked with dots and crosses. The layer B (colour) in Figure 1.2 sits over the hollows marked

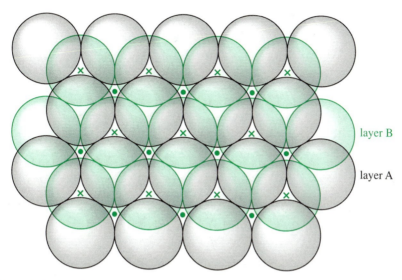

Figure 1.2 Two layers of close-packed spheres.

with a cross (although it makes no difference which type we choose). When we come to add a third layer there are two possible positions where it can go. First, it could go directly over layer A: if we then repeated this stacking sequence we would build up the layers ABABABA ... and so on. This is known as **hexagonal close packing** (**hcp**). Second, the third layer could be positioned over those spaces marked with a dot. This third layer, C, would not be directly over either A or B, and the stacking sequence when repeated would be ABCABCAB ... and so on. This is known as **cubic close packing** (**ccp**). (The names *hexagonal* and *cubic* for these structures arise from the resulting symmetry of the structure; this will be discussed more fully later on.)

Close packing represents the most efficient use of space when packing equal spheres; 74% of the volume is occupied by the spheres and the **packing efficiency** is said to be 74%. Each sphere in the structure is surrounded by 12 equidistant neighbours: six in the same layer, three in the layer above and three in the layer below. The **coordination number** of an atom in a close-packed structure is thus 12.

Another important feature of close-packed structures that we will use extensively later on, is the shape and number of the small amounts of space trapped in between the spheres. There are two different types of space within a close-packed structure: the first we will consider is called an **octahedral hole**. Figure 1.3 shows two close-packed layers again but now with the octahedral holes shaded (in colour). Each of these holes is surrounded by six spheres, three in one layer and three in the other. The centres of the spheres lie at the corners of an octahedron, hence the

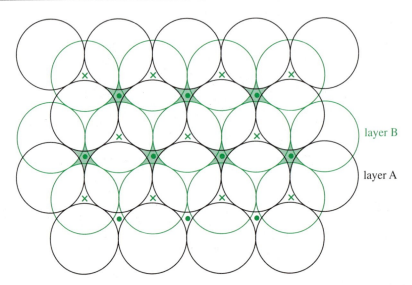

Figure 1.3 Two layers of close-packed spheres with the enclosed octahedral holes picked out in colour.

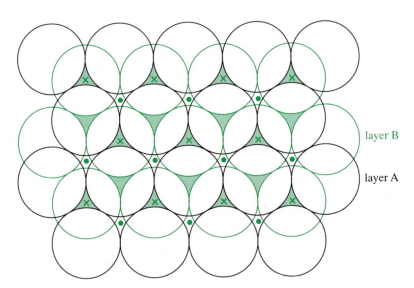

Figure 1.4 Two layers of close-packed spheres with the tetrahedral holes picked out in colour.

name. If there are *n* spheres in the array, then there are also *n* octahedral holes. Similarly, Figure 1.4 shows two close-packed layers, now with the second type of space, **tetrahedral holes**, shaded in colour. Each of these holes is surrounded by four spheres with centres at the corners of a

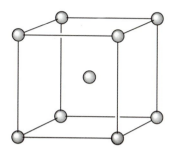

Figure 1.5 A body-centred cubic array.

tetrahedron. If there are n spheres in the array, then there are $2n$ tetrahedral holes. The octahedral holes in a close-packed structure are much bigger than the tetrahedral holes; they are surrounded by six atoms rather than four. It is a matter of simple geometry to calculate that the radius of a sphere that will just fit in an octahedral hole in a close-packed array of spheres of radius r is $0.414r$. For a tetrahedral hole the radius is $0.225r$.

There are of course innumerable stacking sequences that are possible when repeating close-packed layers; however, the hexagonal close-packed and cubic close-packed are those of maximum simplicity and are those most commonly encountered in the crystal structures of the noble gases and of the metallic elements (only two other stacking sequences are found in perfect crystals of the elements: an ABAC repeat in La, Pr, Nd and Am, and a nine-layer repeat ABACACBCB in Sm).

1.3 BODY-CENTRED AND PRIMITIVE STRUCTURES

Some metals do not adopt a close-packed structure but have a slightly less efficient packing method: this is the **body-centred cubic structure (bcc)** shown in Figure 1.5 (Unlike the previous diagrams, the positions of the atoms are represented here, and in subsequent diagrams, by small circles which do not touch: this is merely a device to open up the structure and allow it to be seen more clearly. The whole question of atom and ion size is discussed in section 1.5.4.) In this structure an atom in the middle of a cube is surrounded by eight equidistant atoms at the corners of the cube; the co-ordination number has dropped to eight and the packing efficiency is 68% (compared with 74% for close-packing).

The simplest of the cubic structures is the **primitive cubic structure**. This is built by placing square layers like the one shown in Figure 1.1a, directly on top of one another. Figure 1.6 illustrates this, and you can see

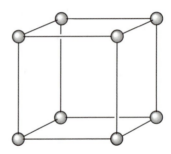

Figure 1.6   A primitive cubic array.

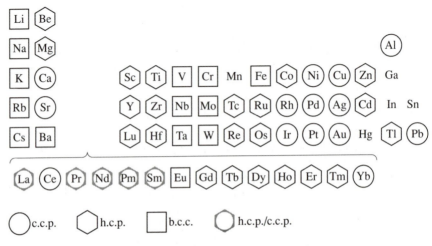

Figure 1.7   Occurrence of packing types among the metals.

that each atom sits at the corner of a cube. The co-ordination number of an atom in this structure is six.

The majority of metals have one of the three basic structures: hcp, ccp, or bcc. Polonium alone adopts the primitive structure. (The bonding in metals is discussed in Chapter 2.) The distribution of the packing types among the most stable forms of the metals at 298 K is shown in Figure 1.7. As we noted earlier, very few metals have a mixed hcp/ccp structure of a more complex type. The structures of the actinides tend to be rather complex and are not included.

1.4 LATTICES AND UNIT CELLS

Crystals are regular-shaped solid particles with flat shiny faces. It was first noted by Hooke in 1664 that the regularity of their external appearance is

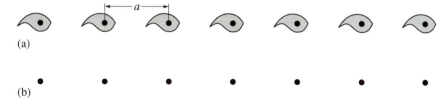

(a)

(b)

Figure 1.8   A one-dimensional lattice.

a reflection of a high degree of internal order. Crystals of the same substance, however, vary in shape considerably. Steno observed in 1671 that this is not because their internal structure varies but because some faces develop more than others. The angle between similar faces on different crystals of the same substance, is always identical. The constancy of the interfacial angles reflects the internal order within the crystals. Each crystal is derived from a basic 'building block' that repeats over and over again, in all directions, in a perfectly regular way. This 'building block' is known as the **unit cell**.

In order to talk about and compare the many thousands of crystal structures that are known, there has to be a way of defining and categorizing the structures. This is achieved by defining the shape and symmetry of each unit cell, and also its size and the positions of the atoms within it.

1.4.1 Lattices

The simplest regular array is a line of evenly spaced objects such as is depicted in Figure 1.8a. There is a dot at the same place in each object: if we now remove the objects leaving the dots, we have a line of equally spaced dots, spacing a (Figure 1.8b). The line of dots is called the **lattice**, and each **lattice point** (dot) must have identical surroundings. This is the only example of a one-dimensional lattice and it can vary only in the spacing a.

1.4.2 Two-dimensional lattices

There are five possible two-dimensional lattices and these are shown in Figure 1.9, together with the restrictions on their repeat distances and angles. Everyday examples of these are to be seen in wallpapers and tiling.

These nets are important in low energy electron diffraction (LEED), which is used to study the structures of surfaces (which are of course two-dimensional) and the pattern of adsorption on a surface of a gas – important in catalysis. In these cases the lattice points are located on

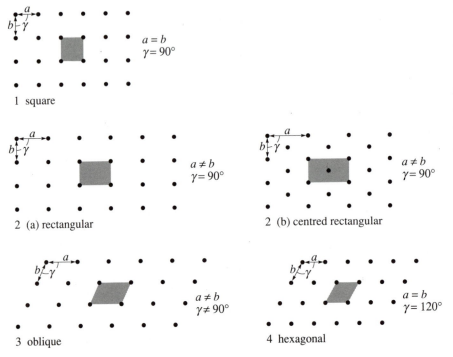

Figure 1.9 The five two-dimensional lattices.

metal atoms, for instance, if studying a metal surface, or on small gaseous molecules such as ammonia, NH_3, if investigating the adsorption of ammonia on a surface.

1.4.3 One and two-dimensional unit cells

The unit cell for the one-dimensional lattice in Figure 1.10a lies between the two vertical lines. If we took this unit cell and repeated it over and over again we would reproduce the original array. Notice that it does not matter where in the structure we place the lattice points as long as they each have identical surroundings. In Figure 1.10b we have moved the lattice points and the unit cell, but repeating this unit cell will still give the same array; we have simply moved the origin of the unit cell. There is never one unique unit cell that is 'correct'; there are always many that could be chosen and the choice depends both on convenience and convention. This is equally true in two and three dimensions.

The conventional unit cells for the five types of two-dimensional lattice are shown in Figure 1.11: notice that they are all parallelograms with their corners at equivalent positions in the array, i.e. the corners of a unit

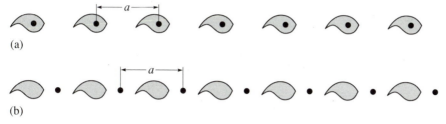

(a)

(b)

Figure 1.10 Choice of unit cell in a one-dimensional lattice.

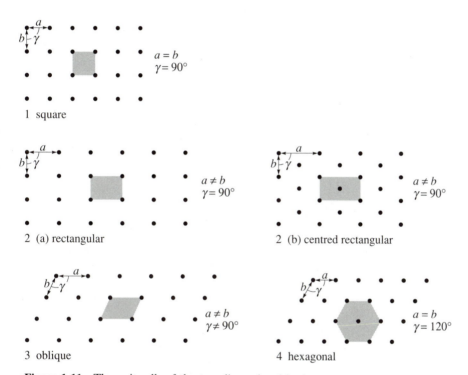

$a = b$
$\gamma = 90°$

1 square

$a \neq b$
$\gamma = 90°$

2 (a) rectangular

$a \neq b$
$\gamma = 90°$

2 (b) centred rectangular

$a \neq b$
$\gamma \neq 90°$

3 oblique

$a = b$
$\gamma = 120°$

4 hexagonal

Figure 1.11 The unit cells of the two-dimensional lattices.

cell are lattice points. The five different unit cell shapes fall into four classes because there are two different types of rectangular unit cell. In Figure 1.12 we show a square array with several different unit cells depicted. All of these, if repeated, would reproduce the array: it is conventional to choose the smallest cell that fully represents the symmetry of the structure. Both unit cells (1a) and (1b) are the same size but clearly (1a) shows that it is a square array, and this would be the conventional choice. Figure 1.13 demonstrates the same principles but for a

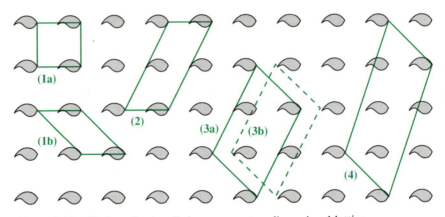

Figure 1.12 Choice of unit cells in a square two-dimensional lattice.

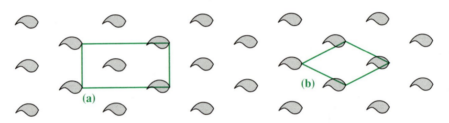

Figure 1.13 Choice of unit cell in a centred rectangular lattice.

centred rectangular array, where (a) would be the conventional choice because it includes information on the centring; the smaller unit cell (b) loses this information. It is always possible to define a non-centred oblique unit cell, but information about the symmetry of the lattice may be lost by doing so.

Unit cells such as (1a) and (1b) in Figure 1.12 and (b) in Figure 1.13 have a lattice point at each corner. However, they each contain a total of one lattice point because each lattice point is shared by four adjacent unit cells. They are known as **primitive unit cells** and given the symbol P. The unit cell marked (a) in figure 1.13 contains a total of two lattice points – one from the shared four corners and one totally enclosed within the cell. This cell is said to be **centred** and is given the symbol C.

1.4.4 Three-dimensional unit cells

When moving into three dimensions matters are more complicated. The unit cell of a three-dimensional lattice is a parallelopiped defined by three distances, a, b and c, and three angles, α, β and γ (Figure 1.14). Of course, because the unit cells are the basic building blocks of the crystals,

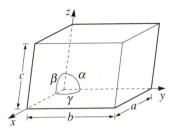

Figure 1.14 Definition of axes, unit cell dimensions and angles for a general unit cell.

Table 1.1 The seven crystal classes

System	Unit cell	Minimum symmetry requirements
Triclinic	$\alpha \neq \beta \neq \gamma \neq 90°$ $a \neq b \neq c$	None
Monoclinic	$\alpha = \gamma = 90°$ $\beta \neq 90°$ $a \neq b \neq c$	One two-fold axis or one symmetry plane
Orthorhombic	$\alpha = \beta = \gamma = 90°$ $a \neq b \neq c$	Any combination of three mutually perpendicular two-fold axes or planes of symmetry
Trigonal	$\alpha = \beta = \gamma \neq 90°$ $a = b = c$	One three-fold axis
Hexagonal	$\alpha = \beta = 90°$ $\gamma = 120°$ $a = b = c$	One six-fold axis or one six-fold improper axis
Tetragonal	$\alpha = \beta = \gamma = 90°$ $a = b \neq c$	One four-fold axis or one four-fold improper axis
Cubic	$\alpha = \beta = \gamma = 90°$ $a = b = c$	Four three-fold axes at 109° 28′ to each other

they must be space-filling, i.e. they must pack together to fill all space. All the possible unit cell shapes are illustrated in Figure 1.15 and their specifications are listed in Table 1.1. These are known as the **seven crystal classes**. The unit cell shapes are determined by minimum symmetry requirements (Table 1.1).

There are four different types of three-dimensional unit cell (Figure 1.16).

1. The **primitive** unit cell (symbol P) contains one lattice point.

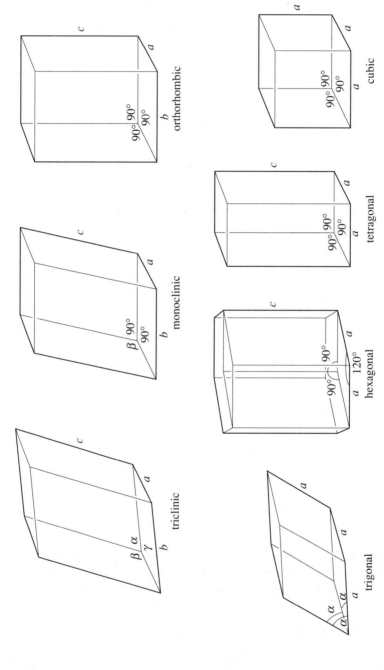

Figure 1.15 The unit cells of the seven crystal systems.

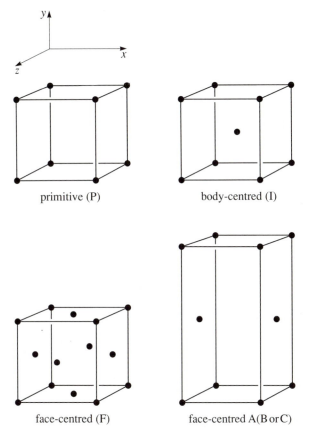

Figure 1.16 Primitive (P), body-centred (I), face-centred (F) and face-centred A, B or C (A shown) unit cells.

2. The **body-centred** unit cell (symbol I) has a lattice point at each corner and also one at the centre of the cell.
3. The **face-centred** unit cell (symbol F) has a lattice point at each corner and one in the centre of each face.
4. The **face-centred** unit cell (symbol A, B or C) has a lattice point at each corner, and one in the centres of one pair of opposite faces, e.g. an A-centred cell has lattice points in the centres of the *bc* faces.

When these four types of lattice are combined with the seven possible unit cell shapes 14 permissible **Bravais lattices** (Figure 1.17) are produced. (It is not possible to combine some of the shapes and lattice types *and* still retain the symmetry requirements listed in Table 1.1. For instance it is not possible to have an A-centred, cubic, unit cell; if only two of the six

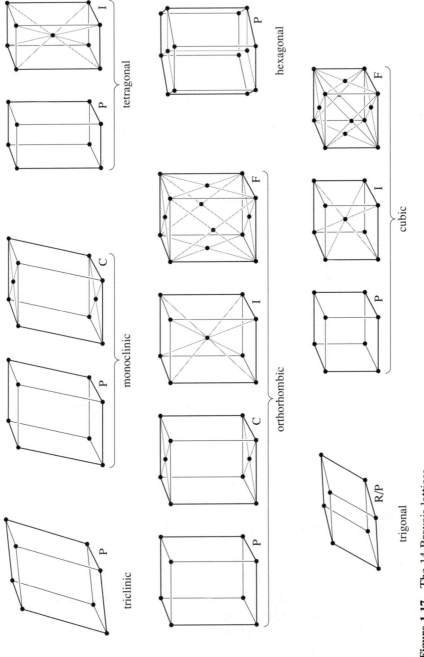

Figure 1.17 The 14 Bravais lattices.

faces are centred, the unit cell necessarily loses its cubic symmetry and cannot have the properties of a cube.)

As a final comment, if we combine all possible symmetry elements with these lattices we find that there are 230 three-dimensional **space groups** that crystal structures can adopt, i.e. 230 different space-filling patterns!

It is important not to lose sight of the fact that the lattice points represent equivalent positions in a crystal structure and not atoms. In a real crystal a lattice point could be occupied by an atom, a complex ion, a molecule or even a group of molecules. The lattice points are used to simplify the repeating patterns within a structure, but they tell us nothing of the chemistry or bonding within the crystal; for that we have to include the atomic positions. This we will do later in the chapter when we look at some real structures.

It is instructive to note how much of a structure these various types of unit cell represent. We noted a difference between the centred and primitive two-dimensional unit cell where the centred cell contains two lattice points whereas the primitive cell contains only one. We can work out similar occupancies for the three-dimensional case. Look back to Figure 1.16 and in each case imagine that the lattice points have been replaced by a single identical molecule. The number of unit cells sharing a particular molecule depends on its site. A corner site is shared by eight unit cells, an edge site by four, a face site by two and a molecule at the body centre is not shared by any other unit cell. Using these figures we can work out the number of molecules in each of the four types of cell in Figure 1.16, assuming that one molecule is occupying each lattice point. The results are listed in Table 1.2.

Table 1.2 Number of molecules in four types of cell

Name	Symbol	Number of molecules in unit cell
Primitive	P	1
Body-centred	I	2
Face-centred	A or B or C	2
All face-centred	F	4

1.4.5 Packing diagrams

Drawing structures in three dimensions is not easy and so crystal structures are often represented by two-dimensional plans or projections of the unit cell contents, in much the same way as an architect makes building plans. These projections are called **packing diagrams** as they are particularly useful in molecular structures for showing how the molecules pack together in the crystal.

The position of an atom or ion in a unit cell is described by its **fractional coordinates**; these are simply the coordinates based on the unit cell axes (known as the **crystallographic axes**), but expressed as fractions of the unit cell lengths. It has the simplicity of a universal system which enables unit cell positions to be compared from structure to structure regardless of variation in unit cell size.

To take a simple example, in a cubic unit cell with $a = 1000$ pm, an atom with an x coordinate of 500 pm has a fractional coordinate in the x direction of $x/a = 500/1000 = 0.5$. Similarly, in the y and z directions, the fractional coordinates are given by y/b and z/c, respectively.

A packing diagram is shown in Figure 1.18 for the body-centred unit cell of Figure 1.5. The projection is shown on the yx plane, i.e. we are looking at the unit cell straight down the z-axis. The z-fractional co-ordinate of any atoms/ions lying in the top or bottom face of the unit cell will be 0 or 1 (depending on where you take the origin) and it is conventional for this not to be marked on the diagram. Any z-coordinate that is not 0 or 1 is marked on the diagram in a convenient place. There is an opportunity to practice constructing these types of diagram in the questions at the end of the chapter.

Many crystallographic papers in the literature contain packing dia-grams. You will notice that often the projection of the unit cell is not taken down an axis on to a unit cell face (for the triclinic system this is not possible anyway as none of the axes are at right-angles). Usually this is because the crystallographer has decided that a better view of the unit cell contents is achieved by rotating the cell somewhat. All of the pro-jections in this book will be taken along a crystallographic axis.

1.5 CRYSTALLINE SOLIDS

We start this section by looking at the structures of some simple **ionic solids**. Ions tend to be formed by the elements in the Groups at the far left and far right of the Periodic Table. Thus we expect the metals in Groups I and II to form cations and the non-metals of Groups VI and VII and nitrogen to form anions, because by doing so they are able to achieve a stable noble gas configuration. Cations can also be formed by some of the Group III elements, such as aluminium (Al^{3+}), by some of the low oxidation state transition metals and even occasionally by the high atomic number elements in Group IV, such as tin and lead, giving Sn^{4+} and Pb^{4+}. Each successive ionization becomes more difficult because the remaining electrons are more strongly bound due to the greater effective nuclear charge, and so highly charged ions are rather rare.

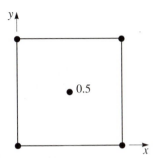

Figure 1.18 Packing diagram for a body-centred unit cell.

An **ionic bond** is formed between two oppositely charged ions because of the electrostatic attraction between them. Ionic bonds are strong but are also non-directional. Ionic crystals are therefore composed of infinite arrays of ions, which have packed together in such a way as to maximize the coulombic attraction between oppositely charged ions and to minimize the repulsions between ions of the same charge. We expect to find ionic compounds in the halides and oxides of the Group I and II metals and it is with these crystal structures that this section begins.

However, just because it is possible to form a particular ion, does not mean that this ion will always exist whatever the circumstances. In many structures we find that the bonding is not purely ionic but possesses some degree of **covalency**: the electrons are shared between the two bonding atoms and not merely transferred from one to the other. This is particularly true for the elements in the centre of the Periodic Table. This point is taken up in section 1.5.4 where we discuss the size of ions and the limitations of the concept of ions as hard spheres.

Two later sections (1.5.5 and 1.5.6) look at the crystalline structures of covalently bonded species. First, extended covalent arrays are investigated, such as the structure of **diamond** (one of the forms of elemental carbon) where each atom forms strong covalent bonds to the surrounding atoms, forming an infinite three-dimensional network of localized bonds throughout the crystal. Second, we look at molecular crystals which are formed from small, individual, covalently bonded molecules. These molecules are held together in the crystal by weak forces known as **van der Waals forces** (or **London dispersion forces**). These forces arise because instantaneous dipoles form in atoms which are in turn capable of inducing other dipoles. The net result is a very weak attractive force that falls off very quickly with distance.

Finally in this section we take a very brief look at the structures of some silicates, the compounds which largely form the earth's crust.

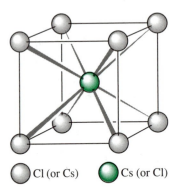

Cl (or Cs)   Cs (or Cl)

Figure 1.19 The crystal structure of caesium chloride (CsCl). This type of diagram is termed a **clinographic projection**, in which the diagram is viewed in a convenient direction from a point at infinity: all parallel lines are still parallel in projection. Important interatomic interactions are represented by heavier lines than the outline of the unit cell. The wedge-shaped bonds are designed to give some feeling of depth to the diagram: the widest part of the wedge is nearest.

1.5.1 Ionic solids with formula MX

The caesium chloride structure

A unit cell of the caesium chloride (CsCl) structure is shown in Figure 1.19. It shows a caesium ion, Cs^+, at the centre of the cubic unit cell, surrounded by eight chloride ions, Cl^-, at the corners. It could equally well have been drawn the other way round with chloride at the centre and caesium at the corners because the structure consists of two inter-penetrating primitive cubic arrays. Note the similarity of this unit cell to the body-centred cubic structure adopted by some of the elemental metals such as the Group 1 (alkali) metals. However, the CsCl structure is *not* body-centred cubic because the environment of the caesium at the centre of the cell is not the same as the environment of the chlorides at the corners: a body-centred cell would have chlorides at the corners, i.e. at (0, 0, 0) etc., *and* at the body centre $(\frac{1}{2}, \frac{1}{2}, \frac{1}{2})$. Each caesium is surrounded by eight chlorines at the corners of a cube and vice versa, so the coordination number of each type of atom is eight. The unit cell contains one formula unit of CsCl, with the eight corner chlorines each being shared by eight unit cells. With ionic structures like this individual molecules are not distinguishable because individual ions are surrounded by ions of the opposite charge.

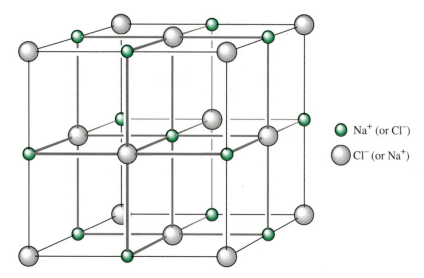

Figure 1.20 The crystal structure of sodium chloride (NaCl).

Caesium is a large ion (ionic radii are discussed in detail in section 1.5.4) and so is able to co-ordinate eight chloride ions around it. Other compounds with large cations that can also accommodate eight anions and crystallize with this structure include CsBr, CsI, TlCl, TlBr, TlI and NH_4Cl.

The sodium chloride (or rock salt) structure

A unit cell of the sodium chloride (NaCl) or rock salt structure is illustrated in Figure 1.20. The unit cell is cubic and the structure consists of two interpenetrating face-centred (F) arrays, one of Na^+ and the other of Cl^- ions. Each sodium ion is surrounded by six equidistant chloride ions situated at the corners of an octahedron and in the same way each chloride ion is surrounded by six sodium ions: we say that the coordination is 6:6.

An alternative way of viewing this structure is to think of it as a cubic close-packed array of chloride ions with sodium ions filling all the octahedral holes. The conventional unit cell of a ccp array is an F face-centred cube (hence the cubic in ccp); the close-packed layers lie at right-angles to a cube diagonal (Figure 1.21). Filling all the octahedral holes gives a Na:Cl ratio of 1:1 with the structure as illustrated in Figure 1.20. Interpreting simple ionic structures in terms of the close packing of one of the ions with the other ion filling some or all of either the octahedral or

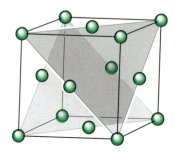

Figure 1.21 A unit cell of sodium chloride showing the position of the close-packed layers.

tetrahedral holes, is extremely useful: it makes it particularly easy to see both the coordination geometry around a specific ion and also the available spaces within a structure.

As you might expect from their relative positions in Group I, a sodium ion is smaller than a caesium ion and so it is now only possible to pack six chlorides around it and not eight as in CsCl. The NaCl unit cell contains four formula units of NaCl. If you find this difficult to see, work it out for yourself by counting up the numbers of ions in the different sites using the information given on p. 15. Table 1.3 lists some of the compounds that adopt the NaCl structure; there are more than 200 known.

Table 1.3 Compounds that have the NaCl (rock salt) type of crystal structure

Most alkali halides, MX, and AgF, AgCl, AgBr
All the alkali hydrides, MH
Monoxides, MO, of Mg, Ca, Sr, Ba
Monosulphides, MS, of Mg, Ca, Sr, Ba

Many of the structures described in this book can be viewed as linked octahedra, where each octahedron consists of a metal atom surrounded by six other atoms situated at the corners of an octahedron (Figure 1.22a, b). These are often depicted, viewed from above with contours marked as in Figure 1.22c. Octahedra can link together via corners, edges and faces, as seen in Figure 1.23. If you find this difficult to visualize, it is quite easy to make small octahedra from paper and see how they join together; Figure 1.24 gives you a template to copy and cut out and glue. (The four shaded triangles provide a template for making tetrahedra.) The linking of octahedra by different methods effectively eliminates atoms because some of the atoms are now shared between them: two

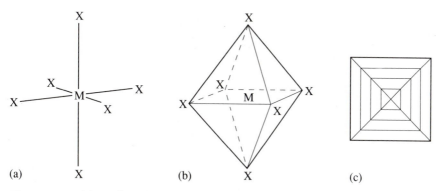

Figure 1.22 (a) An [MX$_6$] octahedron. (b) The solid octahedron. (c) Plan of an octahedron with contours.

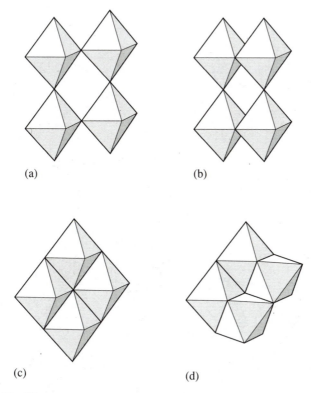

Figure 1.23 The conversion of (a) corner-shared MX$_6$ octahedra to (b) edge-shared octahedra, and (c) edge-shared octahedra to (d) face-shared octahedra.

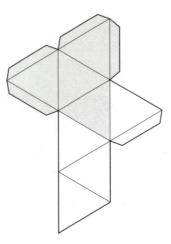

Figure 1.24 Template for making octahedra and tetrahedra.

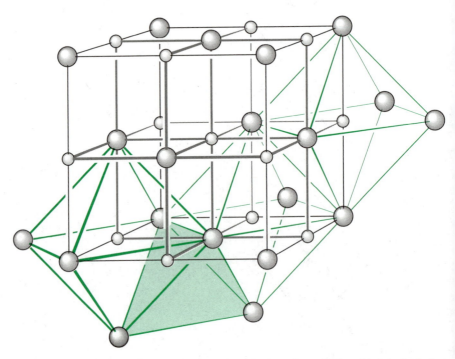

Figure 1.25 NaCl structure showing edge sharing of octahedra. (A tetrahedral space is also shown shaded in colour.)

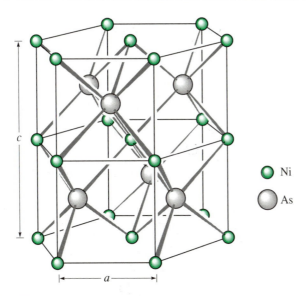

Figure 1.26 The crystal structure of nickel arsenide (NiAs). (For undistorted hcp the ratio $c/a = 1.633$. In reality this ratio varies considerably for the compounds adopting this structure. For example, iron(II) sulphide (FeS) adopts this structure with $c/a = 1.68$.)

$[MO_6]$ octahedra linked through a vertex has the formula, M_2O_{11}; two $[MO_6]$ octahedra linked through an edge has the formula, M_2O_{10}; two $[MO_6]$ octahedra linked through a face has the formula, M_2O_9.

The NaCl structure can be described in terms of $[NaCl_6]$ octahedra sharing edges. An octahedron has 12 edges, and each one is shared by two octahedra in the NaCl structure. This is illustrated in Figure 1.25, which shows a NaCl unit cell with three $[NaCl_6]$ octahedra picked out in colour: the common edges are picked out by a thick broken line. A maximum of only six octahedra can meet at a point, and one of the resulting tetrahedral spaces is depicted by shading.

The nickel arsenide structure

The nickel arsenide (NiAs) structure is the equivalent of the NaCl structure in hexagonal close packing. It can be thought of as an hcp array of arsenic atoms with nickel atoms occupying the octahedral holes (Figure 1.26). The geometry about the nickel atoms is thus octahedral. This is not the case for arsenic: each arsenic atom sits in the centre of a **trigonal prism** of six nickel atoms (Figure 1.27).

The zinc blende and wurtzite structures

Unit cells of these two structures are shown in Figures 1.28 and 1.29, respectively. They are named after two different naturally occurring

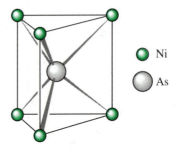

Figure 1.27 The trigonal prismatic coordination of arsenic in NiAs.

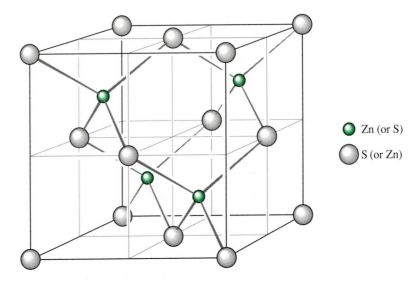

Figure 1.28 The crystal structure of zinc blende (ZnS).

mineral forms of zinc sulphide (ZnS). Structures of the same element or compound which differ only in their atomic arrangements are termed **polymorphs**.

The zinc blende structure can be thought of as a ccp array of sulphide ions with zinc ions occupying every other tetrahedral hole. Each zinc ion is thus tetrahedrally coordinated by four sulphides and vice versa. Compounds adopting this structure include the copper halides and Zn, Cd, and Hg sulphides. Notice that if all the atoms were identical, the structure would be exactly the same as that of diamond.

The wurtzite structure is composed of an hcp array of sulphide ions with alternate tetrahedral holes occupied by zinc ions. Each zinc ion

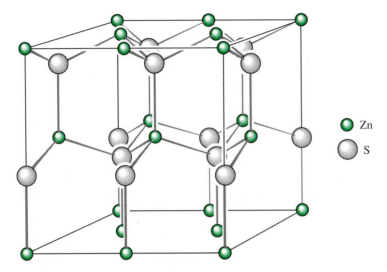

Figure 1.29 The crystal structure of wurtzite (ZnS).

is tetrahedrally coordinated by four sulphide ions and vice versa. Compounds adopting the structure include BeO, ZnO, and NH_4F.

Notice how the coordination numbers of these structures have changed. The coordination number for close packing, where all the atoms are the same size, is 12. In the CsCl structure it is eight, in NaCl, six, and in both of the ZnS structures it is four. As a general rule, the larger a cation is, the more anions it can pack around itself (section 1.5.4).

1.5.2 Solids with general formula MX_2

The fluorite and antifluorite structures

The fluorite structure is named after the mineral form of calcium fluoride (CaF_2) and is illustrated in Figure 1.30. It can be thought of as related to a ccp array of calcium ions with fluorides occupying all the tetrahedral holes. There is a problem with this as a description because calcium ions are rather smaller than fluoride ions, and so physically fluoride ions would not be able to fit into the tetrahedral holes of a calcium ion array. Nevertheless it gives an exact description of the *relative* positions of the ions. The diagram in Figure 1.30a shows the four-fold tetrahedral co-ordination of the fluoride ions very clearly. Notice also that the larger octahedral holes are vacant in this structure; one of them can be seen at the body centre of the unit cell in Figure 1.30a. This becomes a very important feature when we come to look at the movement of ions through defect structures in Chapter 3.

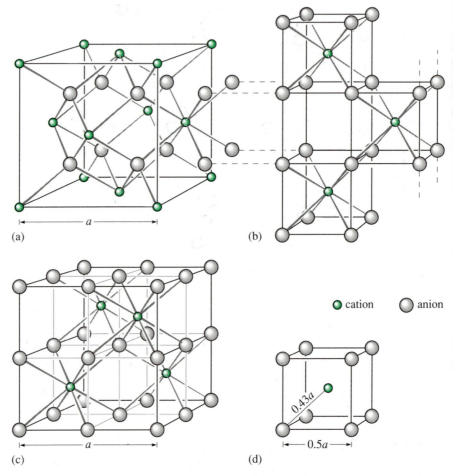

(a)   (b)

cation   anion

(c)   (d)

Figure 1.30 The crystal structure of fluorite (CaF_2). (a) Unit cell as a ccp array of cations. (b) and (c) The same structure redrawn as a primitive cubic array of anions: the unit cell is marked by a coloured outline in (c). (d) Relationship of unit cell dimensions to the primitive anion cube (the octant).

By extending the structure a little and drawing cubes with fluoride ions at each corner, as has been done in Figure 1.30b, you can see that there is eight-fold cubic coordination of each calcium cation. Indeed, it is possible to move the origin and redraw the unit cell so that this feature can be seen more clearly as has been done in Figure 1.30c. The unit cell is now divided into eight smaller cubes called **octants**, with each alternate octant occupied by a calcium cation.

In the antifluorite structure the positions of the cations and anions are merely reversed, and the description of the structure as cations occupying

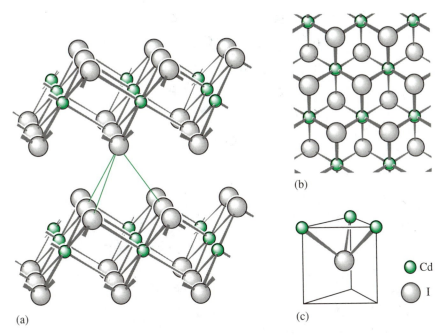

Figure 1.31 The crystal structure of cadmium iodide (CdI_2). (b) The structure of the layers in CdI_2 and $CdCl_2$; the halogen atoms lie in planes above and below that of the metal atoms. (c) The coordination around one iodine atom in CdI_2.

all the tetrahedral holes in a ccp array of anions becomes more realistic. In the example with the biggest anion and smallest cation, Li_2Te, the telluriums are approximately close-packed (even though there is a considerable amount of covalent bonding). For the other compounds adopting this structure such as the oxides and sulphides of the alkali metals, M_2O and M_2S, the description accurately shows the relative positions of the atoms but the anions could not be described as close-packed because they are not touching: the cations are too big to fit in the tetrahedral holes and so the anion–anion distance is greater than for close-packing. These are the only structures where 8:4 coordination is found.

The cadmium chloride and cadmium iodide structures

Both of these structures are based on the close-packing of the appropriate anion with half of the octahedral holes occupied by cations. In both structures the cations occupy all the octahedral holes in every other anion layer giving an overall layer structure with 6:3 coordination. The cadmium chloride ($CdCl_2$) structure is based on a ccp array of chloride ions whereas the cadmium iodide (CdI_2) structure is based on an hcp array of iodide ions. The CdI_2 structure is shown in Figure 1.31. Figure 1.31a shows clearly that an iodide anion in this structure is surrounded by

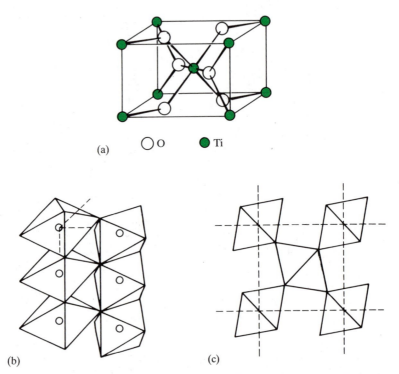

Figure 1.32 The crystal structure of rutile (TiO$_2$). (a) Unit cell; (b) parts of two columns of [TiO$_6$] octahedra; (c) projection of structure on base of unit cell.

three cadmium cations on one side but by three iodides on the other, i.e. it is not completely surrounded by ions of the opposite charge as we would expect for an ionic structure. In fact, this is evidence that the bonding in some of these structures is not entirely ionic as we have tended to imply so far. This is a point that is taken up again in more detail in section 1.5.4.

The rutile structure

The rutile structure is named after one mineral form of titanium dioxide (TiO$_2$). A unit cell is illustrated in Figure 1.32. The unit cell is tetragonal and the structure again demonstrates 6:3 coordination but is not based on close-packing: each titanium atom is coordinated by six oxygens at the corners of a (slightly distorted) octahedron and each oxygen atom is surrounded by three planar titaniums which lie at the corners of an (almost) equilateral triangle. It is not geometrically possible for the co-ordination around titanium to be a perfect octahedron *and* for the co-ordination around oxygen to be a perfect equilateral triangle.

The structure can be viewed as chains of linked $[TiO_6]$ octahedra, where each octahedron shares a pair of opposite edges, and the chains are linked by sharing vertices (Figure 1.32b). Figure 1.32c shows a plan of the unit cell looking down the chains of octahedra so that they are seen in projection.

Occasionally the **antirutile structure** is encountered where the metal and non-metals have swopped places, such as in Ti_2N.

The β-cristobalite structure

The β-cristobalite structure is named after one mineral form of silicon dioxide (SiO_2). The silicon atoms are in the same positions as both the zinc and sulphurs in zinc blende (or the carbons in diamond, which we look at later in section 1.5.5): each pair of silicon atoms is joined by an oxygen midway between. The only metal halide adopting this structure is beryllium fluoride (BeF_2) and it is characterized by 4:2 coordination.

1.5.3 Other important crystal structures

As the valency of the metal increases, the bonding in these simple binary compounds becomes more covalent and the highly symmetrical structures characteristic of the simple ionic compounds occur far less frequently, with molecular and layer structures being common. There are many thousands of inorganic crystal structures; here we describe just a few of those which are commonly encountered and which occur in later chapters.

The bismuth triiodide structure, BiI_3

This structure is based on an hcp array of iodides with the bismuths occupying one-third of the octahedral holes. Alternate pairs of layers have two-thirds of the octahedral sites occupied.

Corundum, α-Al_2O_3

This structure may be described as an hcp array of oxygen atoms with two-thirds of the octahedral holes occupied by aluminium atoms. As we have seen before, geometrical constraints dictate that octahedral co-ordination of the aluminiums precludes tetrahedral coordination of the oxygens. However, it is suggested that this structure is adopted in preference to other possible ones because the four aluminiums surrounding an oxygen approximate most closely to a regular tetrahedron. The structure is also adopted by Ti_2O_3, V_2O_3, Cr_2O_3, α-Fe_2O_3, α-Ga_2O_3 and Rh_2O_3.

The rhenium trioxide structure, ReO_3

This structure (also called the aluminium fluoride structure) is adopted by the fluorides of Al, Sc, Fe, Co, Rh and Pd; also by the oxides WO_3 (at

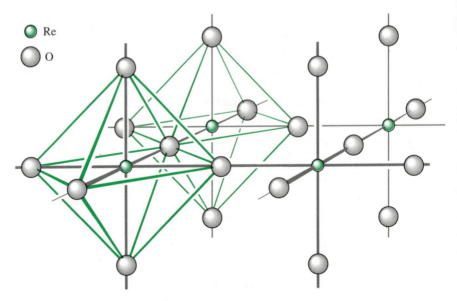

Figure 1.33 Part of the ReO$_3$ structure showing the linking of octahedra through the corners. Part of a single layer.

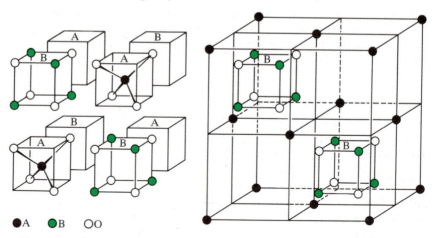

Figure 1.34 The spinel structure, AB$_2$O$_4$.

high temperature) and ReO$_3$ (section 3.8.1). The structure consists of [ReO$_6$] octahedra linked together through each corner to give a highly symmetrical three-dimensional network with cubic symmetry. Part of a layer through the structure is shown in Figure 1.33. Part of two layers is illustrated in Figure 1.35b.

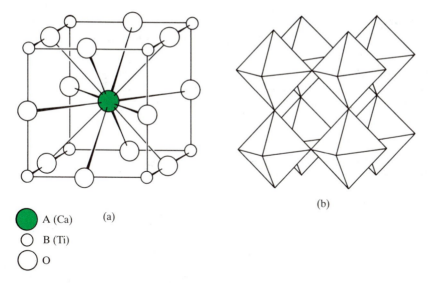

A (Ca)   (a)

B (Ti)

O

Figure 1.35 (a) The perovskite structure for compounds ABO_3, such as $CaTiO_3$. (b) Part of two layers in the ReO_3 structure.

Mixed oxide structures

There are three important mixed oxide structures: **spinel**, **perovskite** and **ilmenite**.

The spinel and inverse-spinel structures

The spinels have the general formula AB_2O_4, taking their name from the mineral spinel $MgAl_2O_4$: generally, A is a divalent ion, A^{2+}, and B is trivalent, B^{3+}. The structure can be thought of as being based on a cubic close-packed array of oxide ions, with A^{2+} ions occupying tetrahedral holes and B^{3+} ions occupying octahedral holes. A spinel crystal containing nAB_2O_4 formula units has $8n$ tetrahedral holes and $4n$ octahedral holes; accordingly one-eighth of the tetrahedral holes are occupied by A^{2+} ions and one-half of the octahedral holes by the B^{3+} ions. A unit cell is illustrated in Figure 1.34; it has been broken down into eight octants, of which there are only two kinds, shown on the left. The A ions occupy tetrahedral positions in the A-type octants together with the corners and face-centres of the unit cell. The B ions occupy octahedral sites, which take up half the corners of the B-type octants. Spinels with this structure include compounds of formula MAl_2O_4 where M is Mg, Fe, Co, Ni, Mn or Zn.

When compounds of general formula AB_2O_4 adopt the inverse-spinel structure, the formula is better written as $B(AB)O_4$, because this indicates that half of the B^{3+} ions now occupy tetrahedral sites, and the remaining half, together with the A^{2+} ions, occupy the octahedral sites. Examples of inverse spinels include Fe_3O_4, $Fe(MgFe)O_4$ and $Fe(ZnFe)O_4$.

The perovskite structure

This structure is named after the mineral $CaTiO_3$. A unit cell is shown in Figure 1.35a: this unit cell is termed the A-type, because if we take the

Table 1.4 Structures related to close-packed arrangements of anions

Formula	Cation: anion coordination	Type and number of holes occupied	Examples	
			Cubic close packing	Hexagonal close packing
MX	6:6	All octahedral	Sodium chloride: NaCl, FeO, MnS, TiC	Nickel arsenide: NiAs, FeS, NiS
	4:4	Half tetrahedral; every alternate site occupied	Zinc blende: ZnS, CuCl, γ-AgI	Wurtzite: ZnS, β-AgI
MX_2	8:4	All tetrahedral;	Fluorite: CaF_2, ThO_2, ZrO_2, CeO_2	None
	6:3	Half octahedral; alternate layers have fully occupied sites	Cadmium chloride: $CdCl_2$	Cadmium iodide: CdI_2, TiS_2
MX_3	6:2	One-third octahedral; alternate pairs of layers have two-thirds of the octahedral sites occupied		Bismuth iodide: BiI_3, $FeCl_3$, $TiCl_3$, VCl_3
M_2X_3	6:4	Two-thirds octahedral		Corundum: Al_2O_3, Fe_2O_3. V_2O_3, Ti_2O_3, Cr_2O_3
ABO_3		Two-thirds octahedral		Ilmenite: $FeTiO_3$
AB_2O_4		One-eighth tetrahedral and one-half octahedral	Spinel: $MgAl_2O_4$ Inverse spinel: $MgFe_2O_4$	Olivine: Mg_2SiO_4

general formula ABX_3 for the perovskites, then in this cell the A atom is in the centre. The central Ca (A) atom is coordinated to eight Ti atoms (B) at the corners and to 12 oxygens (X) at the midpoints of the cell edges. The structure can be usefully described in other ways. First, it can be thought of as a ccp array of A and X atoms with the B atoms occupying the octahedral holes (compare with the unit cell of NaCl in Figure 1.20 if you want to check this). Second, perovskite has the same octahedral framework as ReO_3 based on $[BX_6]$ octahedra with an A atom added in at the centre of the cell (Figure 1.35b). Compounds adopting this structure include $SrTiO_3$, $SrZrO_3$, $SrHfO_3$, $SrSnO_3$ and $BaSnO_3$.

The ilmenite structure
The ilmenite structure is adopted by oxides of formula ABO_3 when A and B are similar in size and their total charge adds up to +6. The structure is named after the mineral of $Fe^{II}Ti^{IV}O_3$, and the structure is very similar to the corundum structure described above, an hcp array of oxygens but now there are two different cations present occupying two-thirds of the octahedral holes. The Fe and Ti atoms occur in alternate layers.

The structures related to close packing are summarized in Table 1.4.

1.5.4 Ionic radii

We know from quantum mechanics that atoms and ions do not have precisely defined radii. However, from the foregoing discussion of ionic crystal structures we have seen that ions pack together in an extremely regular fashion in crystals and that their atomic positions, and thus their interatomic distances, can be measured very accurately. It is a very useful concept, therefore, particularly for those structures based on close-packing, to think of ions as hard spheres, each with a particular radius.

If we take a series of alkali metal halides, all with the rock-salt structure, as we replace one metal ion with another, say sodium with potassium, we would expect the metal-halide internuclear distance to

Table 1.5 Interatomic distances of some alkali halides, r_{M-X}(pm)

	F^-		Cl^-		Br^-		I^-
Li^+	201	56	257	18	275	27	302
	30		24		23		21
Na^+	231	50	281	17	298	25	323
	35		33		31		30
K^+	266	48	314	15	329	24	353
	16		14		14		13
Rb^+	282	46	328	15	343	23	366

change by the same amount each time, if the concept of an ion as a hard sphere with a particular radius holds true. Table 1.5 shows the results of this procedure for a range of alkali halides: the change in internuclear distance on swopping one ion for another is shown in Table 1.5. From this table, we can see that the change in internuclear distance on changing ion is not constant, but also that the variation is not great: this provides us with some experimental evidence that it is not unreasonable to think of the ions as having a fixed radius. We can see, however, that the picture is not precisely true, neither would we expect it to be, because atoms and ions are squashable entities and their size is going to be affected by their environment. Nevertheless, it is a useful concept to develop further as it enables us to describe some of the ionic crystal structures in a simple pictorial way.

There have been many suggestions as to how individual ionic radii can be assigned, and the literature contains several different sets of values: each set is named after the person(s) who originated the method of determining the radii. We will describe some of these methods briefly before listing the values most commonly used at present. It is most important to remember that radii from more than one set of values must not be mixed. Even though the values vary considerably from set to set, each set is internally consistent, i.e. if you add together two radii from one set of values you will obtain an approximately correct internuclear distance as determined from the crystal structure.

The internuclear distances can be determined by X-ray crystallography: in order to obtain values for individual ionic radii from these, the value of one radius needs to be fixed by some method. Originally in 1920, Landé suggested that in the alkali halide with the largest anion and smallest cation, LiI, the iodide ions must be in contact with each other with the tiny Li^+ ion inside the octahedral hole: as the I–I distance is known, it is then a matter of simple geometry to determine the iodide radius. Once the iodide radius is known then the radii of the metal cations can be found from the structures of the metal iodides, and so on. Bragg and Goldschmidt later extended the list of ionic radii using similar methods. It is very difficult to come up with a consistent set of values for the ionic radii, because of course the ions are not hard spheres, they are somewhat elastic, and the radii are affected by their environment, such as the nature of the oppositely charged ligand and the coordination number. Pauling proposed a theoretical method of calculating the radii from the internuclear distances; he produced a set of values that is both internally consistent and also shows the expected trends in the Periodic Table. Pauling's method was to take a series of alkali halides with isoelectronic cations and anions and assume that they are in contact: by assuming that each radius is inversely proportional to the effective nuclear charge felt by the outer electrons of the ion, a radius for each ion can be calculated

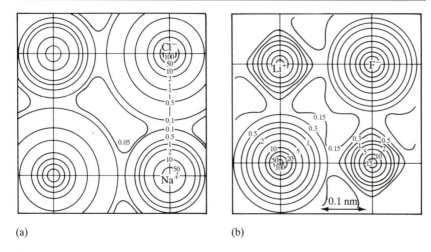

(a)   (b)

Figure 1.36 Electron density maps for (a) NaCl, (b) LiF.

from the internuclear distance. Divalent ions undergo additional compression in a lattice and compensation has to be made for this effect in calculating their radii. With some refinements this method gave a consistent set of values that has been widely used for many years, they are usually known as **effective ionic radii**.

More recently, it has been possible to determine accurate electron density maps for the ionic crystal structures using X-ray crystallography. Such a map is shown for NaCl and for LiF in Figure 1.36. The electron density contours fall to a minimum, although not to zero, in between the nuclei and it is suggested that this minimum position should be taken as the radius position for each ion. These experimentally determined ionic radii are often called **crystal radii**: the values are somewhat different from the older sets and tend to make the anions smaller and the cations bigger than previously; various values for Li^+ and F^- are shown in Figure 1.37. The most comprehensive set of radii has been compiled by Shannon and Prewitt using data from almost 1000 crystal structure determinations. Their crystal radii are based on conventional values of 126 pm and 119 pm for the radii of the O^{2-} and F^- ions, respectively. These values differ by a constant factor of 14 pm from traditional values but it is generally accepted that they correspond more closely to the actual physical size of ions in a crystal. A selection of this data is shown in Table 1.6. Several important trends in the sizes of ions can be noted from the data in Table 1.6.

1. The radii of ions within a Group of the Periodic Table, such as the alkali metals, increase with atomic number, Z. The number of electrons increases as we go down a group, and the outer ones are further from the nucleus.

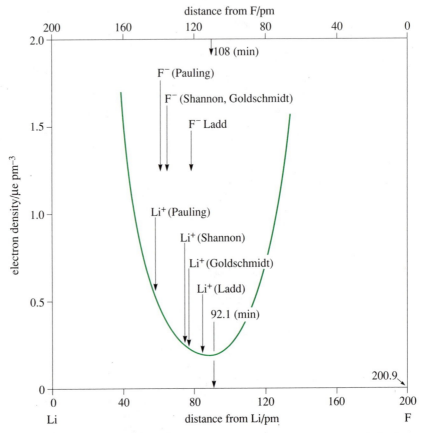

Figure 1.37 Plot showing the variation of electron density along the Li–F direction near the minimum with markers showing the various values of r_+ for Li^+ and r_- for F^- assigned by different methods.

2. In a series of isoelectronic cations, such as Na^+, Mg^{2+}, Al^{3+}, the radius decreases rapidly with increasing positive charge. The number of electrons is constant but the nuclear charge increases and so pulls the electrons in and the radii decrease.

3. For pairs of isoelectronic anions, the radius increases with increasing charge because the more highly charged ion has a smaller nuclear charge.

4. For elements with more than one oxidation state, e.g. Ti^{2+} and Ti^{3+}, the radii decrease as the oxidation state increases; in this case the nuclear charge stays the same but the number of electrons that it acts on decreases.

5. On moving across the Periodic Table for a series of similar ions, such as the first row transition metal divalent ions, M^{2+}, there is an overall

Table 1.6 Crystal radii of some selected ions (pm)*

Ion	Radius	Ion	Radius	Ion	Radius	Ion	Radius	Ion	Radius	Ion	Radius	Ion	Radius
Li^+	90	Be^{2+}	59	B^{3+}	41	C^{4+}	30	Ti^{2+}	100	Ti^{3+}	81	OH^-	123
Na^+	116	Mg^{2+}	86	Al^{3+}	68	Si^{4+}	54	V^{2+}	93	V^{3+}	78	F^-	119
K^+	152	Ca^{2+}	114	Ga^{3+}	76	Ge^{4+}	67	Cr^{2+}	87/94†	Cr^{3+}	76	Cl^-	167
Rb^+	166	Sr^{2+}	132	In^{3+}	94	Sn^{4+}	83	Mn^{2+}	81/97†	Mn^{3+}	72/79†	Br^-	182
Cs^+	181	Ba^{2+}	149	Tl^{3+}	103	Pb^{4+}	92	Fe^{2+}	75/92†	Fe^{3+}	69/79†	I^-	206
								Co^{2+}	79/89†	Co^{3+}	69/75†	O^{2-}	126
Cu^+	74/91‡	Zn^{2+}	74/88‡	Sc^{3+}	89	Ti^{4+}	75	Ni^{2+}	69/63§	Ni^{3+}	70/74†	S^{2-}	170
Ag^+	129	Cd^{2+}	109	Y^{3+}	104	Zr^{4+}	86	Cu^{2+}	87	Cu^{3+}	68/–†	Se^{2-}	184
Au^+	151	Hg^{2+}	116	Lu^{3+}	100	Hf^{4+}	85					$N^{3-‖}$	132
La^{3+}	117	Ce^{3+}	115	Pr^{3+}	113	Nd^{3+}	112	Pm^{3+}	111	Sm^{3+}	110	Eu^{3+}	109
Gd^{3+}	108	Tb^{3+}	106	Dy^{3+}	105	Ho^{3+}	104	Er^{3+}	103	Tm^{3+}	102	Yb^{3+}	101

* Values are taken from R.D. Shannon (1976) *Acta Cryst.*, **A32**, 751, and are for octahedral coordination, unless noted otherwise.
† Low spin/high spin.
‡ Tetrahedral/octahedral.
§ Tetrahedral/square planar.
‖ Tetrahedral.

decrease in radius. This is due to an increase in nuclear charge across the Table because electrons in the same shell do not screen the nucleus from each other very well. A similar effect is seen for the M^{3+} ions of the lanthanides and this is known as the **lanthanide contraction**.

6. For transition metals the spin state affects the ionic radius.
7. The crystal radii increase with an increase in coordination number (see examples Cu^+ and Zn^{2+} in Table 1.6). One can think of fewer ligands around the central ion as allowing the counterions to compress the central ion.

The picture of ions as hard spheres works best for fluorides and oxides, both of which are small, fairly uncompressible ions. As the ions get larger, they are more easily compressed, i.e. the electron cloud is more easily distorted, and they are said to be more **polarizable**.

When discussing particular crystal structures in the previous section it was noted that a larger cation such as Cs^+ was able to pack eight chloride ions around itself whereas the smaller Na^+ only accommodated six. If we continue to think of ions as hard spheres for the present, for a particular structure, as the ratio of the cation and anion radii changes there will come a point when the cation is so small that it will no longer be in touch with the anions. The lattice would not be stable in this state because the negative charges would be too close together for comfort and we would predict that the structure would change to one of lower coordination, allowing the anions to move further apart. If the ions are hard spheres, using simple geometry, it is possible to quantify the radius ratio (i.e. $r_{cation}/r_{anion} = r_+/r_-$) at which this happens, and this is illustrated for the octahedral case in Figure 1.38.

Taking a plane through the centre of an octahedrally coordinated metal cation, the stable situation is shown in Figure 1.38a and the limiting case for stability, when the anions are touching, in Figure 1.38b. The anion radius, r_- in Figure 1.38b is OC, and the cation radius, r_+, is (OA − OC). From the geometry of the right-angled triangle we can see that $\cos 45° = OC/OA = 0.707$. The radius ratio, r_+/r_- is given by (OA − OC)/OC = (OA/OC − 1) = (1.414 − 1) = 0.414. Using similar calculations it is possible to calculate limiting ratios for the other geometries: these are summarized in Table 1.7.

On this basis we would expect to be able to use the ratio of ionic radii to predict possible crystal structures for a given substance. Indeed for many substances this can be done quite successfully: take for example MgO − Mg^{2+} (86 pm), O^{2-} (126 pm) gives $r_+/r_- = 0.68$, correctly predicting six-coordination (MgO has the rock-salt structure). Unfortunately, however, things are not this simple, and taking the radius ratio quite often predicts the wrong structure. In Figure 1.39, where we have plotted the radius ratios of each alkali halide and also indicated what the actual

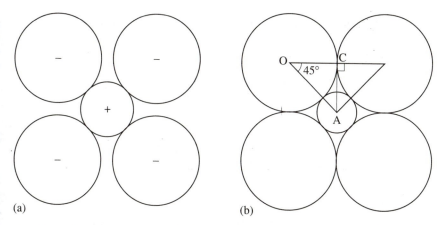

Figure 1.38 (a) Anions packed around a cation on a horizontal plane. (b) Anion–anion contact on a horizontal plane through an octahedron.

Table 1.7 Limiting radius ratios for different coordination numbers

Coordination number	Geometry	Limiting radius ratio	Possible structures
		0.225	
4	Tetrahedral		Wurtzite, zinc blende
		0.414	
6	Octahedral		Rock salt, rutile
		0.732	
8	Cubic		Caesium chloride, fluorite
		1.00	

structure is, only about 50% of the structures are correctly predicted. This plot uses the Shannon and Prewitt radii derived from experimental measurements; however, even if one uses the traditional radii based on the hard-sphere model there is very little improvement in the number of correct predictions.

What is the reason for this? As stated earlier, the model is too simplistic; ions are not hard spheres, but rather are polarized under the influence of cations. This means that the bonding involved is rarely truly ionic but frequently involves at least some degree of covalency. The higher the formal charge on a metal ion the greater will be the proportion of covalent bonding between the metal and its ligands. The higher the degree of covalency, the less likely is the concept of ionic radii and of their ratios to work. It also seems that there is little energy difference between the six-coordinate and eight-coordinate structures and the six-

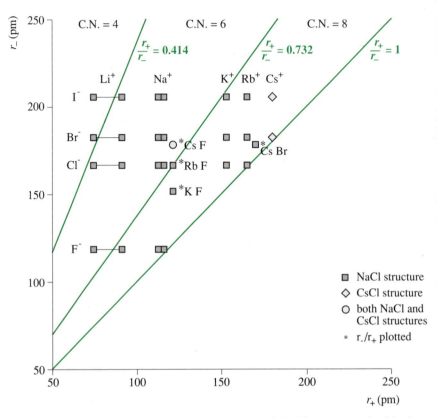

Figure 1.39 Actual crystal structures of the alkali halides compared with the structures predicted by the radius ratio rule. (Because the radii of Li^+ and Na^+ vary significantly with coordination, two values for these ions are given using values of the radii for CN 4 and CN 6.)

coordinate structure is usually preferred; eight-coordinate structures are rarely found, there are no eight-coordinate oxides for instance. The preference for the six-coordinate rock-salt structure is thought to be due to a small amount of covalent-bond contribution: in this structure the three orthogonal p orbitals lie in the same direction as the vectors joining the cation to the surrounding six anions, they are thus well placed for good overlap of the orbitals necessary for σ bonding to take place. The potential overlap of the p orbitals in the caesium chloride structure is less favourable.

1.5.5 Extended covalent arrays

In the last section we noted that many 'ionic' compounds in fact possess some degree of covalency in their bonding. As the formal charge on an

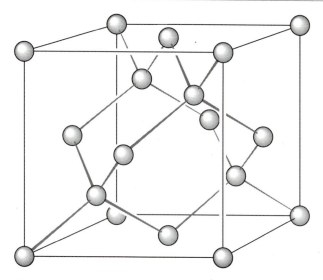

Figure 1.40 A unit cell of the diamond structure.

ion becomes greater we expect the degree of covalency to increase, so we would generally expect compounds of elements in the centre of the Periodic Table to be covalently bonded. Indeed, some of these elements themselves are covalently bonded solids at room temperature. Examples include elements from Group III: boron; Group IV: carbon, silicon, germanium; Group V: phosphorus, arsenic; and Group VI: selenium, tellurium; they form **extended covalent arrays** in their crystal structures.

Take for instance one of the forms of carbon – **diamond**. Diamond has a cubic crystal structure with an F-centred lattice (Figure 1.40); the positions of the atomic centres are the same as in the zinc blende structure, with carbon now occupying both the zinc and the sulphur positions. Each carbon is equivalent and is tetrahedrally linked to four others, forming a **giant molecule** (covalently bonded) throughout the crystal: the carbon–carbon distances are all identical (154 pm). It is interesting to note how the different type of bonding has affected the coordination: here we have identical atoms of the same size, the same situation as for close-packed metals, but the coordination number is now restricted to four because this is the maximum number of covalent bonds that carbon can form. In the case of a metallic element such as magnesium forming a close-packed structure each atom is 12-coordinated (bonding in metals is discussed in Chapter 2). The covalent bonds in diamond are strong and the rigid three-dimensional network of atoms makes diamond the hardest substance known; it also has a high melting temperature (3773 K). Silicon carbide (SiC), known as carborundum, also has this structure, with silicons

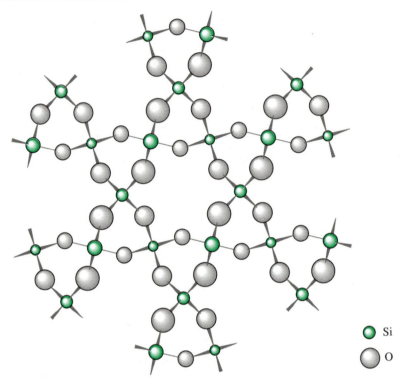

Si

O

Figure 1.41 View down the helical axis of β-quartz.

and carbons alternating throughout; it, too, is very hard and is used for polishing and grinding.

Silica (SiO_2) gives us other examples of giant molecular structures. There are two crystalline forms of silica at atmospheric pressure: **quartz** and **cristobalite**. Each of these also exist in low- and high-temperature forms, α- and β- respectively. (We have already discussed the structure of β-cristobalite in terms of close-packing in section 1.5.2.) Quartz is commonly encountered in nature; the structure of β-quartz is illustrated in Figure 1.41, and consists of $[SiO_4]$ tetrahedra linked so that each oxygen atom is shared by two tetrahedra, thus giving the overall stoichiometry of SiO_2. Notice how once again the covalency of each atom dictates the coordination around itself, silicon having four bonds and oxygen two, rather than the larger coordination numbers that are found for metallic and some ionic structures. The silicon–oxygen bonds in SiO_2 have considerable ionic character. Quartz is unusual in that the linked tetrahedra form **helices** or spirals throughout the crystal, which are all either left- or right-handed, producing laevo- or dextrorotatory crystals, respectively; these are known as **enantiomorphs**.

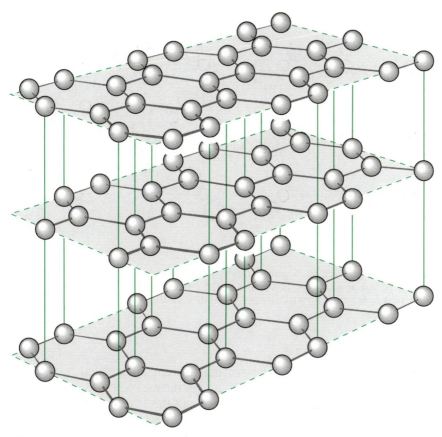

Figure 1.42 The crystal structure of graphite.

For our final example in this section we will look at the structure of normal **graphite**, another polymorph of carbon (Figure 1.42); there are other graphite structures which are more complex. The structure of normal graphite consists of two-dimensional layers of carbon atoms joined together in a hexagonal array. Within the layers each carbon atom is strongly bonded to three others at a distance of 142 pm: this carbon–carbon distance is rather shorter than the one observed in diamond, due to the presence of some double-bonding (the hexagonal configuration of carbon atoms puts some of the *2p* orbitals in a suitable position for π-overlap). The distance between the layers is much greater, 335 pm; this is indicative of weak bonding between the layers due to van der Waals' forces. Graphite is a soft grey solid with a high melting temperature; its softness is attributed to the weak bonding between the layers which allows them to slide over one another.

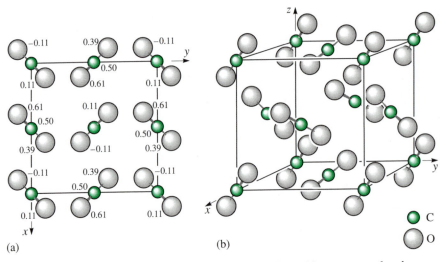

Figure 1.43 (a) Packing diagram of the unit cell of the cubic structure of carbon dioxide (CO_2) projected perpendicular to the z-axis. The heights of the atoms are expressed as fractional coordinates of c. (b) The crystal structure of CO_2.

1.5.6 Molecular structures

Finally we will consider structures that do not contain any extended arrays of atoms. The example of graphite in the previous section in a way forms a bridge between these structures and the structures with infinite three-dimensional arrays. Many crystals contain small, discrete, covalently bonded molecules that are held together only by weak forces, usually either van der Waals' forces or **hydrogen bonding**. A hydrogen bond exists when a hydrogen atom is bonded to two or more other atoms. Usually the hydrogen is attached to two very electronegative atoms such as oxygen or fluorine, with one short covalent bond and one longer, weaker hydrogen bond which can be thought of as a dipole–dipole interaction. Examples of molecular crystals are found throughout organic, organometallic and inorganic chemistry. The crystals are characterized by low melting and boiling temperatures. We will look at just two examples, carbon dioxide and water (ice), both familiar, small, covalently bonded molecules.

Gaseous carbon dioxide (CO_2) when cooled sufficiently forms a molecular crystalline solid (Figure 1.43). Notice that the unit cell contains clearly discernible CO_2 molecules; these are held together in the crystal by weak van der Waals' forces. From structures such as this, atoms can be assigned a **van der Waals' radius**, which is defined as a non-bonded distance of closest approach. Data are compiled from the smallest inter-atomic distances in crystal structures which are not considered to be bonded to one another. If the sum of the van der Waals' radii of two adjacent atoms is greater than the measured distance between them, then it is reasonable to suppose that there is a degree of bonding between them.

The structure of one form of ice (crystalline water) is shown in Figure 1.44 (there are many polymorphs of ice but the one illustrated is the hexagonal form I_h that forms at atmospheric pressure). Each H_2O molecule is tetrahedrally surrounded by four others. The crystal structure is held together by the hydrogen bonds formed between a hydrogen atom on one water molecule and the oxygen of the next, forming a three-dimensional arrangement throughout the crystal.

A summary of the various types of crystalline solids is given in Table 1.8, relating the type of structure to its physical properties. It is important to realize, however, that this only gives a broad overview, intended as a guide only. Not every crystal will fall exactly into one of these categories.

Table 1.8 Classification of crystal structures

Type	Structural unit	Bonding	Characteristics	Examples
Ionic	Cations and anions	Electrostatic, non-directional	Hard, brittle, crystal of high m.t.; moderate insulators; melts are conducting	Alkali metal halides
Extended covalent array	Atoms	Mainly covalent	Strong hard crystal of high m.t.; insulators	Diamond, silica
Molecular	Molecules	Mainly covalent between atoms in molecule, van der Waals' or hydrogen bonding between molecules	Soft crystal of low m.t. and large coefficient of expansion; insulators	Ice, organic compounds
Metallic	Metal atoms	Band model (see Chapter 2)	Single crystals are soft; strength depends on structural defects and grain; good conductors; m.t.s. vary but tend to be high	Iron, aluminium, sodium

m.t. = Melting temperature.

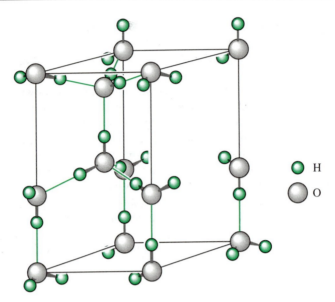

Figure 1.44   The crystal structure of hexagonal ice (I_h).

1.5.7 Silicates

The **silicates** form a large group of crystalline compounds with rather complex but interesting structures. A great part of the earth's crust is formed from these complex oxides of silicon.

Silicon itself crystallizes with the same structure as diamond. Its normal oxide, silica (SiO_2), is polymorphic and in previous sections we have discussed the crystal structure of two of its polymorphs: β-cristobalite (section 1.5.2) and β-quartz (section 1.5.5). Quartz is one of the commonest minerals on the earth, occurring as sand on the seashore, as a constituent in granite and flint and, in less pure form, as agate and opal. The silicon atom in all these structures is tetrahedrally coordinated.

The silicate structures are most conveniently discussed in terms of the $[SiO_4]^{4-}$ unit. The $[SiO_4]^{4-}$ unit has tetrahedral coordination of silicon by oxygen and is represented in these structures by a small tetrahedron as shown in Figure 1.45a. The silicon–oxygen bonds possess considerable covalent character.

There are some minerals, **olivines** for instance, which contain discrete $[SiO_4]^{4-}$ tetrahedra: these compounds do not contain Si—O—Si—O—Si—... chains but there is considerable covalent character in the metal–silicate bonds. These are often described as **orthosilicates**, salts of orthosilicic acid, $Si(OH)_4$ or H_4SiO_4, which is a very weak acid. The structure of olivine itself, $(Mg,Fe)_2SiO_4$, which can be described as an

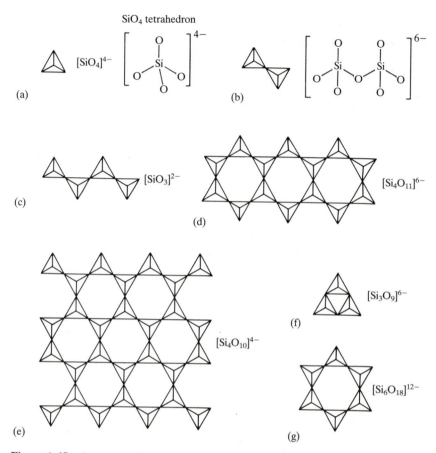

Figure 1.45 A structural classification of mineral silicates.

assembly of $[SiO_4]^{4-}$ ions and Mg^{2+} (or Fe^{2+}) ions, appears earlier in Table 1.4 because an alternative description is of an hcp array of oxygens with silicons occupying one-eighth of the tetrahedral holes and magnesium ions occupying one-half of the octahedral holes.

In most silicates, however, the $[SiO_4]^{4-}$ tetrahedra are linked by oxygen sharing through a vertex, such as is illustrated in Figure 1.45b for two linked tetrahedra to give $[Si_2O_7]^{6-}$. Notice that each terminal oxygen confers a negative charge on the anion and the shared oxygen is neutral. The diagrams of silicate structures showing the silicate frameworks (Figure 1.45) omit these charges as they can be readily calculated. By the sharing of one or more oxygen atoms through the vertices, the tetrahedra are able to link up to form chains, rings, layers, etc. The negative charges on the silicate framework are balanced by metal cations in the lattice. Some examples are discussed below.

Discrete $[SiO_4]^{4-}$ units

Examples are found in **olivine** (see above), an important constituent of basalt; **garnets**, $M_3^{II}M_2^{III}(SiO_4)_3$ (where M^{II} can be Ca^{2+}, Mg^{2+}, or Fe^{2+} and M^{III} can be Al^{3+}, Cr^{3+} or Fe^{3+}), the framework of which is composed of $[M^{III}O_6]$ octahedra which are joined to six others via vertex-sharing $[SiO_4]$ tetrahedra, the M^{II} ions are coordinated by eight oxygens in dodecahedral interstices; Ca_2SiO_4, found in mortars and Portland cement; **zircon** ($ZrSiO_4$) which has eight-coordinate Zr.

Disilicate units, $[Si_2O_7]^{6-}$

Structures containing this unit (Figure 1.45b) are not common but occur in **thortveitite** ($Sc_2Si_2O_7$) and **hemimorphite** ($Zn_4(OH)_2Si_2O_7$).

Chains

$[SiO_4]^{4-}$ units share two corners to form infinite chains (Figure 1.45c). The repeat unit is $[SiO_3]^{2-}$. Minerals with this structure are called **pyroxenes**, e.g. **diopside** ($CaMg(SiO_3)_2$) and **enstatite** ($MgSiO_3$). The silicate chains lie parallel to one another and are linked together by the cations which lie between them.

Double chains

Here alternate tetrahedra share two and three oxygen atoms, respectively (Figure 1.45d). This class of minerals is known as the **amphiboles**, an example of which is **tremolite** ($Ca_2Mg_5(OH)_2(Si_4O_{11})_2$). Most of the asbestos minerals fall in this class. The repeat unit is $[Si_4O_{11}]^{6-}$.

Infinite layers

The tetrahedra all share three oxygen atoms (Figure 1.45e). The repeat unit is $[Si_4O_{10}]^{4-}$. Examples are **talc** ($Mg_3(OH)_2Si_4O_{10}$) which contains a sandwich of two layers with octahedrally coordinated Mg^{2+} between the layers; **micas**, e.g. $KMg_3(OH)_2Si_3AlO_{10}$; and clay minerals such as **kaolin** ($Al_2(OH)_4Si_2O_5$).

Rings

Each $[SiO_4]^{4-}$ unit shares two corners as in the chains. Figures 1.45f and g show three and six tetrahedra linked together; these have the general formula $[SiO_3]_n^{-2n}$; rings also may be made from four tetrahedra. An example of a six-tetrahedra ring is **beryl** (emerald; $Be_3Al_2Si_6O_{18}$); here the rings lie parallel with metal ions between them. Other examples include **dioptase** ($Cu_6Si_6O_{18}.6H_2O$) and **benitoite** ($BaTiSi_3O_9$).

Three-dimensional structures

If $[SiO_4]^{4-}$ tetrahedra share all four oxygens, then the structure of silica (SiO_2) is produced. However, if some of the silicon atoms are replaced by the similarly sized atoms of the Group III element aluminium (i.e. if $[SiO_4]^{4-}$ is replaced by $[AlO_4]^{5-}$), then other cations must be introduced to balance the charges. Such minerals include the **feldspars** (general formula, $M(Al,Si)_4O_8$) the most abundant of the rock-forming minerals; the **zeolites**, which are used as ion exchangers, molecular sieves and catalysts (these are discussed in detail in Chapter 5); the **ultramarines**, which are coloured silicates manufactured for use as pigments, **lapis lazuli** being a naturally occurring mineral of this type.

As one might expect there is an approximate correlation between the solid state structure and the physical properties of a particular silicate. For instance, cement contains discrete $[SiO_4]^{4-}$ units and is soft and crumbly; asbestos minerals contain double chains of $[SiO_4]^{4-}$ units and are characteristically fibrous; mica contains infinite layers of $[SiO_4]^{4-}$ units and the weak bonding between the layers is easily broken and micas show cleavage parallel to the layers; granite contains feldspars which are based on three-dimensional $[SiO_4]^{4-}$ frameworks and are very hard.

1.6 LATTICE ENERGY

The **lattice energy (L_0)** of a crystal is defined as the internal energy change when one mole of the solid is formed from the gaseous ions at infinite separation, at atmospheric pressure and 0 K. For example, in sodium chloride we can approximate this to be the enthalpy change for the reaction in equation (1.1) under these conditions.

$$Na^+(g) + Cl^-(g) = NaCl(s) \tag{1.1}$$

As it is not possible to measure lattice energies directly, they are usually determined either from a **Born–Haber cycle** or by calculation.

1.6.1 The Born–Haber cycle

A Born–Haber cycle is the application of **Hess's law** to the enthalpy of formation of an ionic solid at 298 K. Hess's law states that the enthalpy of a reaction is the same whether the reaction takes place in one step or in several. A Born–Haber cycle for a metal chloride, MCl, is shown in Figure 1.46; the metal chloride is formed from the constituent elements in their standard state in the equation at the bottom, and by the clockwise series of steps above. From Hess's law, the sum of the enthalpy changes for each step around the cycle can be equated with the standard enthalpy of formation, giving:

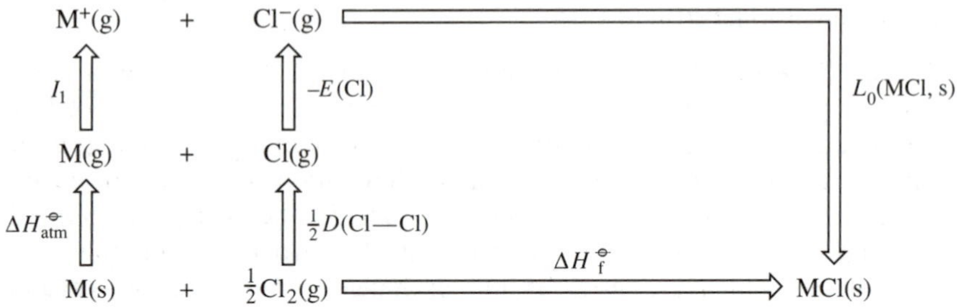

Figure 1.46   The Born–Haber cycle for a metal chloride, MCl.

$$\Delta H_f^{\ominus}(\text{MCl,s}) = \Delta H_{\text{atm}}^{\ominus}(\text{M,s}) + I_1(\text{M}) + \tfrac{1}{2}D_m(\text{Cl–Cl})$$
$$- E(\text{Cl}) + L_0(\text{MCl,s}). \tag{1.2}$$

By rearranging this equation, we can write an expression for the lattice energy in terms of the other quantities, which can then be calculated if the values for these are known. The terms in the Born–Haber cycle are defined in Table 1.9 together with some sample data. Notice that the way in which we have defined lattice energy gives **negative** values; equation (1.1) may be written the other way round in some texts, in which case

Table 1.9   Terms in the Born–Haber cycle

Term	Definition of the reaction to which the term applies	NaCl (kJ mol^{-1})	AgCl (kJ mol^{-1})
$\Delta H_{\text{atm}}^{\ominus}(\text{M})$	M(s) = M(g) standard enthalpy of atomization of metal M	107.8	284.6
$I_1(\text{M})^*$	M(g) = M$^+$(g) + e$^-$(g) first ionization energy of metal M	494	732
$\tfrac{1}{2}D(\text{Cl–Cl})^*$	$\tfrac{1}{2}$Cl$_2$(g) = Cl(g) half the dissociation energy of Cl$_2$	122	122
$-E(\text{Cl})^*$	Cl(g) + e$^-$(g) = Cl$^-$(g) the energy change of this reaction is defined as minus the electron affinity of chlorine	−349	−349
$L_0(\text{MCl,s})^*$	M$^+$(g) + Cl$^-$(g) = MCl(s) lattice energy of MCl(s)		
$\Delta H_f^{\ominus}(\text{MCl,s})$	M(s) + $\tfrac{1}{2}$Cl$_2$(g) = MCl(s) standard enthalpy of formation of MCl(s)	−411.1	−127.1

*These quantities are defined as internal energy changes at 0 K. In practice this is approximated by using the enthalpy change for the reaction shown.

positive lattice energies will be quoted. It is not important which convention is chosen, providing you stick to it! The energy changes for each step are standard enthalpies, and although the symbol L_0 is conventionally used in the cycle for lattice energy, strictly we should have used L_{298} because this is the temperature at which standard enthalpies are determined. The error introduced into the determination of L_0 by this approximation is only of the order of $10 \, kJ \, mol^{-1}$.

Cycles such as this can clearly be constructed for other compounds such as oxides, MO, sulphides, MS, higher valent metal halides, MX_n, etc. The difficulty in these cycles sometimes comes in the determination of values for the electron affinity, E. (We can think of E as being the heat evolved when an electron is added to an atom: as an enthalpy change refers to the heat absorbed, the two must have opposite signs.) In the case of an oxide, it is necessary to know the double electron affinity for oxygen (minus the enthalpy change of the following reaction):

$$2e^-(g) + O(g) = O^{2-}(g). \tag{1.3}$$

This can be broken down into two stages:

$$e^-(g) + O(g) = O^-(g) \tag{1.4}$$

and

$$e^-(g) + O^-(g) = O^{2-}(g). \tag{1.5}$$

It is impossible to determine the enthalpy of reaction for equation (1.5) experimentally, and so this value can only be found if the lattice energy is known – a Catch 22 situation! To overcome problems such as this, methods of calculating (rather than measuring) the lattice energy have been devised and they are described in the next section.

1.6.2 Calculating lattice energies

For an ionic crystal of known structure, it should be a simple matter to calculate the energy released on bringing the ions from an infinite separation to form the crystal, using the equations of simple electrostatics. The energy of an ion pair, $M^+ X^-$ (assuming they are point charges), separated by distance, r, is given by **Coulomb's law**:

$$E = -\frac{e^2}{4\pi\varepsilon_0 r} \tag{1.6}$$

and where the *magnitudes* of the charges on the ions are Z_+ and Z_-, for the cation and anion respectively, by

$$E = -\frac{Z_+ Z_- e^2}{4\pi\varepsilon_0 r} \tag{1.7}$$

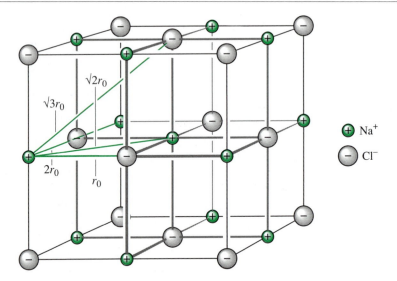

Figure 1.47   Sodium chloride structure showing internuclear distances.

(e is the electronic charge, $1.6 \times 10^{-19}\,C$, and ε_0 is the permittivity of a vacuum, $8.854 \times 10^{-12}\,F\,m^{-1}$)

The energy due to coulombic interactions in a crystal is calculated for a particular structure by summing all the ion pair interactions, thus producing an infinite series. The series will include terms due to the attraction of the opposite charges on cations and anions and also repulsion terms due to cation–cation and anion–anion interactions. Some of these interactions are shown in Figure 1.47 for the NaCl structure: the Na^+ ion in the centre is immediately surrounded by six Cl^- ions at a distance of r, then by 12 cations at a distance of $\sqrt{2}r$, then by eight anions at $\sqrt{3}r$, followed by a further six cations at $2r$, and so on. The coulombic energy of interaction E_C is given by the summation of all these interactions:

$$E_C = -\frac{e^2}{4\pi\varepsilon_0 r}\left(6 - \frac{12}{\sqrt{2}} + \frac{8}{\sqrt{3}} - \frac{6}{2} + \frac{24}{\sqrt{5}}\cdots\right). \tag{1.8}$$

The term inside the parentheses is known as the **Madelung constant (A)**, in this case for the NaCl structure. The series is slow to converge but nevertheless values of the Madelung constant have been computed, not only for NaCl, but for most of the simple ionic structures. For one mole of NaCl we can write:

$$E_C = -\frac{N_A A e^2}{4\pi\varepsilon_0 r}, \tag{1.9}$$

where N_A is the **Avogadro number**, $6.022 \times 10^{23}\,mol^{-1}$. (Note that the expression is multiplied by N_A and not by $2N_A$, even though there are N_A

cations and N_A anions present; this avoids counting every interaction twice!) The value of the Madelung constant is dependent only on the geometry of the lattice, and not on its dimensions; values for various structures are given in Table 1.10.

Table 1.10 Madelung constants for some common ionic lattices

Structure	Madelung constant (A)	$\dfrac{A}{v}$	Coordination
Caesium chloride	1.763	0.88	8:8
Sodium chloride	1.748	0.87	6:6
Fluorite	2.519	0.84	8:4
Zinc blende	1.638	0.82	4:4
Wurtzite	1.641	0.82	4:4
Corundum	4.172	0.83	6:4
Rutile	2.408	0.80	6:3

Ions, of course, are *not* point charges, but consist of positively charged nuclei surrounded by electron clouds. At small distances these electron clouds repel each other, and this too needs to be taken into account when calculating the lattice energy of the crystal. At large distances the repulsion energy is negligible but as the ions approach one another closely, it increases very rapidly. Born suggested that the form of this repulsive interaction could be expressed by:

$$E_R = \frac{B}{r^n},\tag{1.10}$$

where B is a constant and n (known as the **Born exponent**) is large and also a constant.

We can now write an expression for the lattice energy in terms of the energies of the interactions that we have considered:

$$L_0 = E_C + E_R = -\frac{N_A A Z_+ Z_- e^2}{4\pi\varepsilon_0 r} + \frac{B}{r^n}.\tag{1.11}$$

The lattice energy will be a minimum when the crystal is at equilibrium, i.e. when the internuclear distance is at the equilibrium value of r_0. We can minimize the lattice energy function by differentiating with respect to r and equating to zero:

$$\frac{dL_0}{dr} = \frac{N_A A Z_+ Z_- e^2}{4\pi\varepsilon_0 r^2} - \frac{nB}{r^{n+1}}$$

but

$$\frac{dL_0}{dr} = 0, \quad \text{when } r = r_0$$

$$\frac{nB}{r_0^{n+1}} = \frac{N_A A Z_+ Z_- e^2}{4\pi\varepsilon_0 r_0^2}$$

$$B = \frac{N_A A Z_+ Z_- e^2 r_0^{n+1}}{4\pi\varepsilon_0 r_0^2} = \frac{N_A A Z_+ Z_- e^2 r_0^{n-1}}{4\pi\varepsilon_0}$$

$$L_0 = -\frac{N_A A Z_+ Z_- e^2}{4\pi\varepsilon_0 r_0} + \frac{N_A A Z_+ Z_- e^2 r_0^{n-1}}{4\pi\varepsilon_0 r_0^n}$$

$$L_0 = -\frac{N_A A Z_+ Z_- e^2}{4\pi\varepsilon_0 r_0}\left(1 - \frac{1}{n}\right). \tag{1.12}$$

This is known as the **Born–Landé equation**: the values of r_0 and n can be obtained from X-ray crystallography and from compressibility measurements, respectively. The other terms in the equation are well-known constants, and when values for these are substituted we get:

$$\boxed{L_0 = -\frac{1.389 \times 10^5 \, A Z_+ Z_-}{r_0}\left(1 - \frac{1}{n}\right).} \tag{1.13}$$

If the units of r_0 are pm, then the units of L_0 will be kJ mol^{-1}.

Pauling showed that the values of n can be approximated with reasonable accuracy for compounds of ions with noble gas configurations, by averaging empirical constants for each ion. The values of these constants are given in Table 1.11. For example, n for rubidium chloride (RbCl) is 9.5 (average of 9 and 10) and for strontium chloride (SrCl$_2$) is 9.33 (average of 9, 9 and 10).

Table 1.11 Constants used to calculate n

Ion type	Constant
[He]	5
[Ne]	7
[Ar]	9
[Kr]	10
[Xe]	12

It has been suggested that the repulsion term in the lattice energy expression is better represented by:

$$E_R = b e^{-(r/\rho)} \tag{1.14}$$

b and ρ are constants also determined from compressibility measurements. This gives an expression for the lattice energy of:

$$L_0 = -\frac{N_A A Z_+ Z_- e^2}{4\pi\varepsilon_0 r_0}\left(1 - \frac{\rho}{r_0}\right).$$ (1.15)

This is known as the **Born–Mayer equation**.

Notice what a dramatic effect the charge on the ions can have on the value of the lattice energy. A structure containing one doubly charged ion has a factor of two in the equation ($Z_+ Z_- = 2$), whereas one containing two doubly charged ions is multiplied by a factor of four ($Z_+ Z_- = 4$). Structures containing multiply charged ions tend to have much larger (numerically) lattice energies.

Some lattice energy values that have been calculated by various methods are shown in Table 1.12 for comparison with experimental values that have been computed using a Born–Haber cycle. Remarkably good agreement is achieved, considering all the approximations involved. The largest discrepancies are for the large polarizable ions, where, of course, the ionic model is not expected to be perfect. The equations can be improved to help with these discrepancies by including the effect of van der Waals' forces, zero point energy (the energy due to the vibration of the ions at

Table 1.12 Lattice energies of some alkali and alkaline earth metal halides at 0 K

Compound	Structure	$L_0(kJ\,mol^{-1})$			
		Born–Haber cycle*	Born–Landé equation (1.12)[†]	Extended calculation[‡]	Kapustinskii equation (1.17)
LiF	NaCl	−1025		−1033	−970
LiI	NaCl	−756		−738	−725
NaF	NaCl	−910	−904	−906	−882
NaCl	NaCl	−772	−757	−770	−753
NaBr	NaCl	−736	−720	−735	−720
NaI	NaCl	−701	−674	−687	−673
KCl	NaCl	−704	−690	−702	−679
KI	NaCl	−646	−623	−636	−613
CsF	NaCl	−741	−724	−734	−716
CsCl	CsCl	−652	−623	−636	−629
CsI	CsCl	−611	−569	−592	−572
MgF$_2$	Rutile	−2922	−2883	−2914	
CaF$_2$	Fluorite	−2597	−2594	−2610	
CaCl$_2$	Deformed rutile	−2226		−2223	

* D.A. Johnson (1982) *Some Thermodynamic Aspects of Inorganic Chemistry*, Cambridge.
[†] D.F.C. Morris (1957) *J. Inorg. Nucl. Chem*, **4**, 8.
[‡] D. Cubiociotti (1961) *J. Chem. Phys.*, **34**, 2189; T.E. Brackett and E.B. Brackett (1965) *J. Phys. Chem.*, **69**, 3611; H.D.B. Jenkins and K.F. Pratt (1977) *Proc. R. Soc. Series A*, **356**, 115.

0 K) and heat capacity. The net effect of these corrections is only of the order of $10 \, kJ \, mol^{-1}$ and the values thus obtained for the lattice energy are known as **extended-calculation values**.

It is important to note that the good agreement achieved between the Born–Haber and calculated values for lattice energy, do not in any way prove that the ionic model is valid. This is because the equations possess a self-compensating feature in that they use *formal charges* on the ions, but take *experimental internuclear distances*. The values of r_0 are the result of all the various types of bonding in the crystal, not just of the ionic bonding, and so are rather shorter than one would expect for purely ionic bonding.

Kapustinskii noted that if the Madelung constants, A, for a number of structures are divided by the number of ions in one formula unit of the structure, v, the resulting values are almost constant (Table 1.10), and this led to the idea that it would be possible to set up a general lattice energy equation that could be applied to any crystal regardless of its structure. For the six-coordinate NaCl structure, the value of A/v is 0.874, and this can be plugged into either the Born–Landé or the Born–Mayer equation and used to calculate the lattice energy of an unknown structure. In these equations r_0 can then be replaced by $(r_+ + r_-)$, where r_+ and r_- are the cation and anion radii for six-coordination; n can be assigned an average value of 9. With these substitutions in equation (1.12) we get:

$$L_0 = -\frac{1.079 \times 10^5 \, vZ_+Z_-}{r_+ + r_-}.$$

(1.16)

ρ in equation (1.15) is approximately 34.5 pm for the alkali halides, and so substituting in equation (1.15) gives us:

$$L_0 = -\frac{1.214 \times 10^5 \, vZ_+Z_-}{r_+ + r_-}\left(1 - \frac{34.5}{r_+ + r_-}\right).$$

(1.17)

These are known as the **Kapustinskii equations**. Values computed using equation (1.17) are also contained in Table 1.12 for comparison with the other equations.

As we end this section let us reconsider ionic radii briefly. Many ionic compounds contain complex or polyatomic ions. Clearly, it is going to be extremely difficult to measure the radii of ions such as ammonium (NH_4^+) or carbonate (CO_3^{2-}), for instance. However, Yatsimirskii has devised a method which determines a value of the radius of a polyatomic ion by applying the Kapustinskii equation to lattice energies determined from thermochemical cycles. Such values are called **thermochemical radii** and Table 1.13 lists some of these values.

Table 1.13 Thermochemical radii of polyatomic ions*

Ion	pm	Ion	pm	Ion	pm
NH_4^+	151	ClO_4^-	226	MnO_4^{2-}	215
Me_4N^+	215	CN^-	177	O_2^{2-}	144
PH_4^+	171	CNS^-	199	OH^-	119
$AlCl_4^-$	281	CO_3^{2-}	164	PtF_6^{2-}	282
BF_4^-	218	IO_3^-	108	$PtCl_6^{2-}$	299
BH_4^-	179	N_3^-	181	$PtBr_6^{2-}$	328
BrO_3^-	140	NCO^-	189	PtI_6^{2-}	328
CH_3COO^-	148	NO_2^-	178	SO_4^{2-}	244
ClO_3^-	157	NO_3^-	165	SeO_4^{2-}	235

* J.E. Huheey (1983) *Inorganic Chemistry*, 3rd edn, Harper and Row, London, based on data from H.D.B. Jenkins and K.P. Thakur (1979) *J. Chem. Ed.*, **56**, 576.

1.6.3 Calculations using thermochemical cycles and lattice energies

It is not yet possible to measure a lattice energy directly, which is why the best experimental values for the alkali halides (Table 1.12) are derived from a thermochemical cycle. This in itself is not always easy for compounds other than the alkali halides, because as noted before, not all of the data are necessarily available. Electron affinity values are known from experimental measurements for most of the elements, but when calculating a lattice energy for a sulphide (say) then we need to know the enthalpy change for the reaction in equation (1.18)

$$2e^-(g) + S(g) = S^{2-}(g) \tag{1.18}$$

which is minus a **double electron affinity** and is not so readily available. This is where lattice energy calculations come into their own, because we can use one of the methods discussed above to calculate a value of L_0 for the appropriate sulphide and then plug it into the thermochemical cycle to calculate the enthalpy change of equation (1.18).

Proton affinities can be accessed in a similar way: a proton affinity is defined as the enthalpy change of the reaction shown in equation (1.19), where a proton is lost by the species, A.

$$AH^+(g) = A(g) + H^+(g). \tag{1.19}$$

This value can be obtained from a suitable thermochemical cycle providing the lattice energy is known. Take as an example the formation of the ammonium ion, $NH_4^+(g)$:

$$NH_3(g) + H^+(g) = NH_4^+(g). \tag{1.20}$$

The enthalpy change of the reaction in equation (1.20) is minus the proton affinity of ammonia, $-P(NH_3(g))$. This could be calculated from

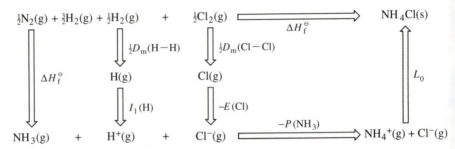

Figure 1.48 Thermochemical cycle for the calculation of the proton affinity of ammonia.

the thermochemical cycle shown in Figure 1.48, provided the lattice energy of ammonium chloride is known.

Thermochemical cycles can also be used to provide us with information on the thermodynamic properties of compounds with metals in unusual oxidation states which have not yet been prepared. For instance, we can use arguments to determine whether it is possible to prepare a compound of sodium in a higher oxidation state than normal, $NaCl_2$. In order to calculate a value for the enthalpy of formation and thence $\Delta G_f^{\ominus}(NaCl_2(s))$, we need to set up a Born–Haber cycle for $NaCl_2(s)$ of the type shown in Figure 1.49. The only term in this cycle that is not known is the lattice energy of $NaCl_2$, and for this we make the approximation that it will be the same as that of $MgCl_2$ which is known. The summation becomes:

$$\Delta H_f^{\ominus}(NaCl_2(s)) = \Delta H_{atm}^{\ominus}(Na(s)) + I_1(Na) + I_2(Na) + D_m(Cl\text{–}Cl)$$
$$- 2E(Cl) + L_0(NaCl_2(s)). \qquad (1.21)$$

Table 1.14 Values of the Born–Haber cycle terms for $NaCl_2$ and $MgCl_2$ ($kJ\,mol^{-1}$)

	Na	Mg
$\Delta H_{atm}^{\ominus}$	108	148
I_1	494	736
I_2^{*}	4565	1452
$D(Cl\text{–}Cl)$	244	244
$-2E(Cl)$	-698	-698
$L_0(MCl_2(s))$	-2523	-2523
$\Delta H_f^{\ominus}(MCl_2(s))$	2190	-641

*The second ionization energy, I_2, refers to the energy change of the reaction: $M^+(g) - e^-(g) = M^{2+}(g)$.

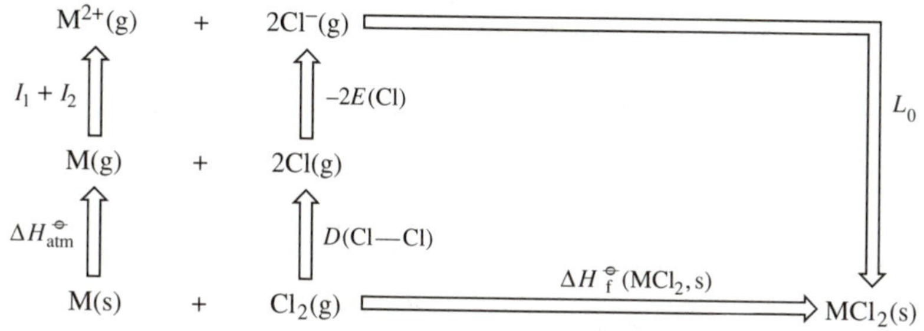

Figure 1.49 Born–Haber cycle for a metal dichloride, MCl_2.

We noted earlier that solids containing doubly charged ions tend to have large negative lattice energies, yet the calculation for $NaCl_2$ has shown just the opposite. Why? A glance at the figures in Table 1.14 reveals the answer: the second ionization energy for sodium is huge (because this relates to the removal of an inner shell electron) and this far outweighs the large lattice energy. So, $NaCl_2$ does not exist because the extra stabilization of the lattice due to a doubly charged ion is not enough to compensate for the large second ionization energy.

The calculation we have just performed shows that for the reaction

$$NaCl_2(s) = Na(s) + Cl_2(g): \Delta H_m^\ominus = -2190 \, kJ \, mol^{-1}.$$

To be sure of the stability of the compound we need a value for $\Delta G_m^\ominus$. For the analogous reaction of $MgCl_2(s)$, $\Delta S_m^\ominus = 166.1 \, J \, K^{-1} \, mol^{-1}$, and so

$$\Delta G_m^\ominus = \Delta H_m^\ominus - T\Delta S_m^\ominus = -2190 - (298.15 \times 0.166)$$
$$= -2239 \, kJ \, mol^{-1},$$

suggesting that $NaCl_2$ would be unstable with respect to the elements sodium and chlorine.

Interestingly it was arguments and calculations of this sort that led Neil Bartlett to the discovery of the first noble gas compound, $XePtF_6$. Bartlett had prepared a new complex, O_2PtF_6, which by analogy with the diffraction pattern of $KPtF_6$, he formulated as containing the dioxygenyl cation, $[O_2^+][PtF_6^-]$. He realized that the ionization energies of oxygen and xenon are very similar and that although the radius of the Xe^+ ion is slightly different, because the PtF_6^- anion is very large the lattice energy of $[Xe^+][PtF_6^-]$ should be very similar to that of the dioxygenyl complex and therefore should exist! Accordingly he mixed xenon and PtF_6 and obtained the orange-yellow solid of xenon hexafluoroplatinate – the first noble gas compound. (At room temperature the $XePtF_6$ reacts with another molecule of PtF_6 to give a product containing $[XeF]^+[PtF_6]^-$ and $[PtF_5]^-$.)

Several examples of problems involving these types of calculation are included in the questions at the end of the chapter.

1.7 CONCLUSION

In this opening chapter we have introduced many of the principles and ideas that lay behind a discussion of the crystalline solid state. We have discussed in detail the structure of a number of important ionic crystal structures and shown how they can be linked to a simple view of ions as hard spheres that pack together as closely as possible, but can also be viewed as the linking of octahedra or tetrahedra in various ways. Taking these ideas further we have investigated the size of these ions in terms of their radii and thence the energy involved in forming a lattice with ionic bonding. We also noted that covalent bonding is present in many structures and that when covalent bonding only is present we tend to see a rather different type of crystal structure.

Chapter 2 looks at a more modern approach to the bonding in solids – the band model.

FURTHER READING

Wells, A.F. (1984) *Structural Inorganic Chemistry*, 5th edn, Oxford University Press, Oxford.
   This is the classic and comprehensive reference text for inorganic crystal structures.
Evans, R.C. (1966) *An Introduction to Crystal Chemistry*, Cambridge University Press, Cambridge.
   A traditional treatment of crystal structures at the undergraduate level.
Adams, D.M. (1974) *Inorganic Solids*, Wiley, New York.
   Well written text, accessible to undergraduates.
Johnson, D.A. (1982) *Some Thermodynamic Aspects of Inorganic Chemistry*, 2nd edn, Cambridge University Press, Cambridge.
   Detailed discussion of the thermodynamics of inorganic solids. Suitable for higher level undergraduates and postgraduates.
Huheey, J.E. (1983) *Inorganic Chemistry*, 3rd edn, Harper and Row, London.
   Excellent, comprehensive undergraduate text, with a good discussion of structure and bonding.

QUESTIONS

1. How many tetrahedral and octahedral holes are there in a ccp array of *n* spheres?

2. What is the coordination number of any atom in an infinite ccp array?

3. What is the coordination number of an atom in an infinite primitive array?

4. Assuming an asymmetric object, e.g. a molecule, to be placed on each lattice point, evaluate the number of molecules in a unit cell, for each type of unit cell, i.e. P, I, F and A, and see if they agree with the results in Table 1.2.

5. How many centred cells are drawn in Figure 1.12?

6. What are the coordination numbers of Cs^+ and Cl^- in the caesium chloride structure?

7. What is the Bravais lattice of the caesium chloride structure and how many formula units of CsCl are there in the unit cell?

8. What are the coordination numbers of Na^+ and Cl^- in the rock-salt structure?

9. Calculate the number of formula units of NaCl in the unit cell illustrated in Figure 1.20.

10. Describe and draw the geometry around Ni and As in the NiAs structure. What is the coordination number of each type of atom?

11. What are the coordination numbers of Zn and S in the zinc blende structure?

12. Using Figures 1.19, 1.20 and 1.28 together with models if necessary, draw unit cell projections for (a) CsCl; (b) NaCl; (c) ZnS (zinc blende).

13. How many formula units, ZnS, are there in the zinc blende unit cell?

14. Draw a projection of a unit cell for both the hcp and ccp structures, seen perpendicular to the close-packed layers (i.e. assume that the close-packed layer is the *ab* plane, draw in the *x*- and *y*-coordinates of the atoms in their correct positions and mark the third coordinate *z* as a fraction of the corresponding repeat distance *c*).

15. Estimate a value for the radius of the iodide ion. The distance between the lithium and iodine nuclei in lithium iodide is 300 pm.

16. Calculate a radius for F^- from the data in Table 1.5 for NaI and NaF. Repeat the calculation using RbI and RbF.

17. Use the Born–Haber cycle in Figure 1.49 and the data in Table 1.15 to calculate the lattice energy of solid calcium chloride ($CaCl_2$).

18. Calculate the value of the Madelung constant for the structure in Figure 1.50. All bond lengths are equal and all bond angles are 90°. Assume that there are no ions other than those shown in the figure

Table 1.15 Values of the Born–Haber cycle terms for $CaCl_2$ ($kJ\,mol^{-1}$)

Term	Value
$\Delta H^{\ominus}_{atm}$	178
I_1	590
I_2	1146
$D(Cl-Cl)$	244
$-2E(Cl)$	−698
$\Delta H^{\ominus}_f(CaCl_2(s))$	−795.8

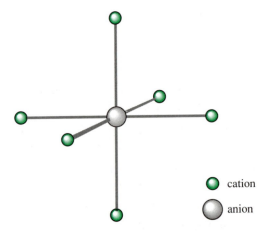

cation

anion

Figure 1.50 Structure for use with question 18. All bond lengths are equal and bond angles are 90°.

and that the charges on the cations and anions are +1 and −1, respectively.

19. Calculate a value for the lattice energy of potassium chloride using equation (1.17). Compare this with the value you calculate from the thermodynamic data in Table 1.16.

Table 1.16 Values of the Born–Haber cycle terms for KCl ($kJ\,mol^{-1}$)

Term	Value
$\Delta H^{\ominus}_{atm}$	89.1
I_1	418
$\frac{1}{2}D(Cl-Cl)$	122
$-E(Cl)$	−349
$\Delta H^{\ominus}_f(KCl(s))$	−436.7

20. Calculate a value for the electron affinity of sulphur for two electrons. Take iron(II) sulphide (FeS) as a model and devise a suitable cycle. Use the data given in Table 1.17.

Table 1.17 Values of the Born–Haber cycle terms for FeS (kJ mol^{-1})

Term	Value
$\Delta H_{atm}^{\ominus}(Fe(s))$	416.3
$I_1(Fe)$	761
$I_2(Fe)$	1561
$\Delta H_{atm}^{\ominus}(S(s))$	278.8
$\Delta H_f^{\ominus}(FeS(s))$	−100.0
$E(S)$	200

21. Calculate a value for the electron affinity of oxygen for two electrons. Take magnesium oxide (MgO) as a model and devise a suitable cycle. Use the data given in Table 1.18.

Table 1.18 Values of the Born–Haber cycle terms for MgO (kJ mol^{-1})

Term	Value
$\Delta H_{atm}^{\ominus}(Mg(s))$	147.7
$I_1(Mg)$	736
$I_2(Mg)$	1452
$\frac{1}{2}D_m(O-O)$	249
$\Delta H_f^{\ominus}(MgO(s))$	−601.7
$E(O)$	141

22. Calculate a value for the proton affinity of ammonia using the cycle in Figure 1.48 and data in Table 1.19.

Table 1.19 Values of the Born–Haber cycle terms for NH$_4$Cl (kJ mol^{-1})

Term	Value
$\Delta H_f^{\ominus}(NH_3)$	−46.0
$\Delta H_f^{\ominus}(NH_4Cl(s))$	−314.4
$\frac{1}{2}D_m(H-H)$	218
$\frac{1}{2}D_m(Cl-Cl)$	122
$I(H)$	1314
$E(Cl)$	349
$r_+(NH_4^+)$	151 pm

23. Compounds of aluminium and magnesium in the lower oxidation states, Al(I) and Mg(I), do not exist under normal conditions. If we make an assumption that the radius of Al^+ or Mg^+ is the same as that of Na^+ (same row of the periodic table) then we can also equate the lattice energies, MCl. Use this information in a Born–Haber cycle to calculate a value of the enthalpy of formation, $\Delta H_f^{\ominus}$, for AlCl(s) and MgCl(s), using the data in Table 1.20.

Table 1.20 Values of the Born–Haber cycle terms for NaCl, MgCl and AlCl ($kJ\,mol^{-1}$)

Term	Na	Mg	Al
$\Delta H_{atm}^{\ominus}$	108	148	326
I_1	494	736	577
$\frac{1}{2}D(Cl-Cl)$	122	122	122
$-E(Cl)$	-349	-349	-349
$\Delta H_f^{\ominus}(NaCl(s))$	-411		
$L_0(MCl(s))$			

ANSWERS

1. There are $2n$ tetrahedral holes and n octahedral holes; the same is true for an hcp array.

2. Twelve. Again, this is the same for a hcp array.

3. The coordination number of each atom is six.

4. P: the corner of each unit cell is shared with eight other unit cells. The unit cell has eight corners and, therefore, contains one molecule.
   I: the body-centred cell has ($8 \times \frac{1}{8}$) molecules at the corners, and one molecule in the centre which is not shared, making two molecules in all.
   A: the A-centred cell has only two faces centred, contributing ($2 \times \frac{1}{2}$) molecules, and ($8 \times \frac{1}{8}$) at the corners, making a total of two molecules.
   F: each face is shared by two unit cells, therefore the face-centred cell has ($6 \times \frac{1}{2}$) molecules on the faces and ($8 \times \frac{1}{8}$) at the corners, making a total of four molecules.

5. There are four. Unit cell (2) contains two lattice points; (3a) and (3b) each contain three lattice points; (4) contains four lattice points.

6. Both Cs^+ and Cl^- have a coordination number of eight, as each ion is surrounded by eight equivalent oppositely charged ions at the corners of the cube.

7. CsCl has a primitive lattice. If you find this answer difficult to understand and think that CsCl is body-centred, look back to the definition of a lattice on p. 7. This structure would be body-centred only if the ion in the centre of the cell were the same as the ions at the corners.

   There is one formula unit of CsCl in the unit cell. This is made up from one Cl^- in the centre of the cell, plus eight Cs^+ ions at the corners which all contribute one-eighth of a Cs^+ ion to the cell, as they are shared by eight adjacent unit cells.

8. The coordination number is six for both Na^+ and Cl^- ions; each ion is octahedrally coordinated by the other.

9. There are four Na^+ ions: eight at cube corners shared with eight unit cells, and six at cube faces each shared with two unit cells, making a total of $(8 \times \frac{1}{8}) + (6 \times \frac{1}{2}) = 4$. There are four Cl^- ions: 12 at midpoints of cube edges shared with four unit cells, one unshared at the body-centre position, making a total of $(12 \times \frac{1}{4}) + 1 = 4$. Thus, there are four formula units of NaCl in the cell.

10. The Ni atoms are octahedrally coordinated by six As atoms. The As atoms sit in the centre of a trigonal prism of six Ni atoms. The coordination number of both As and Ni is six.

11. The coordination number is four in each case. Each Zn atom is surrounded tetrahedrally by four S atoms and vice versa. This is also true for the wurtzite structure.

12. (a) The unit cell projection for CsCl is shown in Figure 1.51a. (b) The unit cell projection for NaCl is shown in Figure 1.51b. This is a view looking down c on the unit cell, showing ions at level 0.5 in c in black and at levels 0 and 1 in c in colour. (c) The unit cell projection for ZnS (zinc blende) is shown in Figure 1.51c.

13. Four. If the sulphur atoms in Figure 1.28 are black circles, there are $(6 \times \frac{1}{2}) = 3$ at the centres of the faces and $(8 \times \frac{1}{8}) = 1$ at the corners. These four are matched by the four zincs entirely enclosed in the cell.

14. (a) Figure 1.52a shows the hexagonal unit cell for hcp. (b) Figure 1.52b shows a possible unit cell for ccp, which has three-fold symmetry. A more usual unit cell for this structure is a cubic one which is more difficult to visualize because the close-packed planes lie parallel to a body diagonal of the cell (see Figure 1.21).

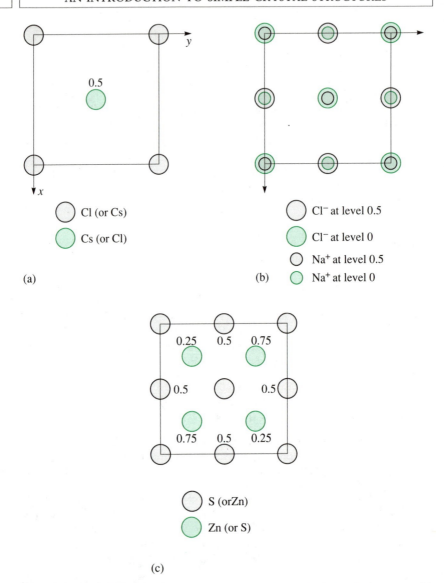

Cl (or Cs)

Cs (or Cl)

(a)

Cl⁻ at level 0.5

Cl⁻ at level 0

Na⁺ at level 0.5

Na⁺ at level 0

(b)

S (orZn)

Zn (or S)

(c)

Figure 1.51 (a) Plan of the CsCl structure. (b) Plan of the NaCl structure. (c) Plan of the zinc blende, ZnS, structure.

15. Assuming that anion–anion contact occurs as in Figure 1.38b, the iodide ion radius is $300/\sqrt{2}$ or 212 pm.

16. From the internuclear distance in NaI, $r(Na^+) = (323 - 212)$ pm $= 111$ pm. Then from the internuclear distance in NaF, $r(F^-) =$

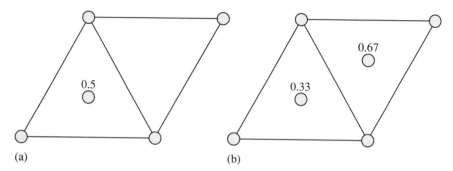

Figure 1.52 (a) Hexagonal close packing. (b) Cubic close packing.

$(231 - 111)\,pm = 120\,pm$. The same procedure for RbI and RbF gives $r(Rb^+) = 154\,pm$ and $r(F^-) = 128\,pm$.

17. $L_0(CaCl_2(s)) = -2255.8\,kJ\,mol^{-1}$. From equation (1.21) we can write

$$\Delta H_f^{\ominus}(CaCl_2(s)) = \Delta H_{atm}^{\ominus}(Ca(s)) + I_1(Ca) + I_2(Ca)$$
$$+ D_m(Cl-Cl) - 2E(Cl) + L_0(CaCl_2(s)).$$

Rearranging and inserting the values from Table 1.15 gives the answer. The lattice energy for $CaCl_2$ has a large negative value. Note how dominant in the cycle is the second ionization energy of calcium.

18. The Madelung constant, $A = -3.99$. There are seven ions in the structure in Figure 1.50: six cations and one anion. First calculate the contribution to the potential energy of interactions of the six cations with the central anion. Each cation is distance r_0 from the central ion.

$$E_1 = -\frac{6e^2}{4\pi\varepsilon_0 r_0}.$$

Each cation also interacts with a diametrically opposite cation (distance $2r_0$). There are three such interactions, so

$$E_2 = +\frac{3e^2}{4\pi\varepsilon_0 2r_0}.$$

Finally, E_3 is calculated from interactions between adjacent cations (distance $\sqrt{2}r_0$) of which there are 12

$$E_3 = +\frac{12e^2}{4\pi\varepsilon_0 \sqrt{2}r_0}$$

$$E = E_1 + E_2 + E_3$$

$$= -\frac{e^2}{4\pi\varepsilon_0 r_0}\left(6 - \frac{3}{2} - \frac{12}{\sqrt{2}}\right)$$

$$= -\frac{Ae^2}{4\pi\varepsilon_0 r_0}$$

so,

$$A = 6 - \frac{3}{2} - \frac{12}{\sqrt{2}}$$

$$A = -3.99.$$

19. The appropriate Born–Haber cycle is shown in Figure 1.46 and leads to the relationship in equation (1.2):

$$\Delta H_f^{\ominus}(\mathrm{MCl(s)}) = \Delta H_{\mathrm{atm}}^{\ominus}(\mathrm{M(s)}) + I_1(\mathrm{M}) + \tfrac{1}{2}D_{\mathrm{m}}(\mathrm{Cl\text{-}Cl}) - E(\mathrm{Cl})$$
$$+ L_0(\mathrm{MCl(s)}).$$

Using the data in Table 1.16 we have

$$L_0 = (-436.7 - 89.1 - 418 - 122 + 349)\,\mathrm{kJ\,mol}^{-1}$$
$$= -717\,\mathrm{kJ\,mol}^{-1}.$$

From equation (1.17):

$$L_0 = -\frac{1.214 \times 10^5 v Z_+ Z_-}{r_+ + r_-}\left(1 - \frac{34.5}{r_+ + r_-}\right),$$

for KCl, $Z_+ = 1$, $Z_- = 1$, $r_{K^+} = 152\,\mathrm{pm}$, $r_{Cl^-} = 167\,\mathrm{pm}$, $v = 2$, giving $L_0 = -679\,\mathrm{kJ\,mol}^{-1}$. There is a discrepancy of $38\,\mathrm{kJ\,mol}^{-1}$ between the calculated value and the thermodynamic value.

20. An appropriate cycle is shown in Figure 1.53. Note that sulphur is a solid in its standard state not a gas. We wish to calculate $[E(\mathrm{S}) + E(\mathrm{S}^-)]$, which we can do if $L_0(\mathrm{FeS(s)})$ can be calculated. Values of all

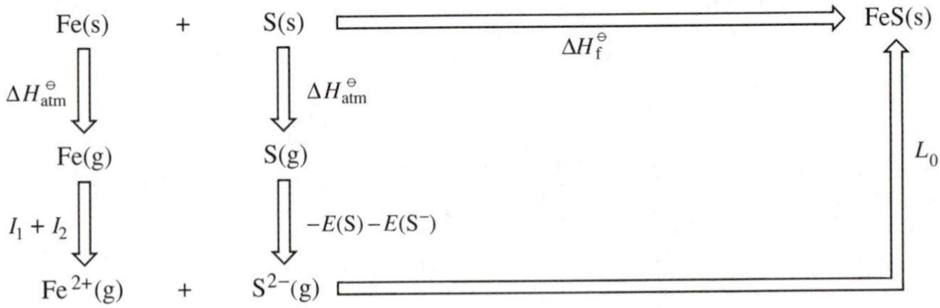

Figure 1.53 Born–Haber cycle for the calculation of the electron affinity of sulphur.

the other terms in the cycle are known. The lattice energy relationship of equation (1.17) will be needed:

$$L_0 = -\frac{1.214 \times 10^5 v Z_+ Z_-}{r_+ + r_-}\left(1 - \frac{34.5}{r_+ + r_-}\right).$$

Substituting $v = 2$, $Z_+ = 2$, $Z_- = 2$, $r_+ = 92$, $r_- = 170$, gives $L_0 = -3219\,\text{kJ mol}^{-1}$. From the cycle in Figure 1.53

$$E(S) + E(S^-) = -\Delta H_f^{\ominus}(\text{FeS(s)}) + \Delta H_{atm}^{\ominus}(\text{Fe(s)})$$
$$+ (I_1 + I_2) + \Delta H_{atm}^{\ominus}(\text{S(s)})$$
$$+ L_0(\text{FeS(s)})$$
$$= (100.0 + 416.3 + 761 + 1561 + 278.8$$
$$- 3219)$$
$$= -101.9\,\text{kJ mol}^{-1}$$
$$E(S) = 200, \text{ so } E(S^-) = -101.9 - 200 = -301.9\,\text{kJ mol}^{-1}.$$

The value of the electron affinity is the heat given out on the addition of an electron, so this implies that the enthalpy change for the addition of an electron to the $S^-(g)$ anion is endothermic:

$$S^-(g) + e^-(g) = S^{2-}(g); \qquad \Delta H_m^{\ominus} = 301.9\,\text{kJ mol}^{-1}.$$

Not surprisingly it appears to be energetically unfavourable to add an electron to a negatively charged ion.

21. An appropriate cycle is shown in Figure 1.54. We wish to calculate $[E(O) + E(O^-)]$, which we can do if $L_0(\text{MgO(s)})$ can be calculated. Values of all the other terms in the cycle are known. The lattice energy relationship of equation (1.17) will be needed:

$$L_0 = -\frac{1.214 \times 10^5 v Z_+ Z_-}{r_+ + r_-}\left(1 - \frac{34.5}{r_+ + r_-}\right).$$

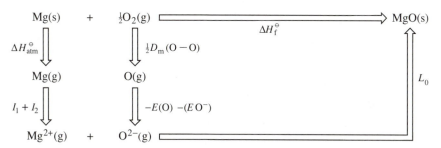

Figure 1.54 Born–Haber cycle for magnesium oxide.

Substituting $v = 2$, $Z_+ = 2$, $Z_- = 2$, $r_+ = 86$, $r_- = 126$, gives $L_0 = -3836\,\text{kJ}\,\text{mol}^{-1}$. From the cycle in Figure 1.54

$$
\begin{aligned}
E(O) + E(O^-) &= -\Delta H_f^{\ominus}(\text{MgO(s)}) + \Delta H_{\text{atm}}^{\ominus}(\text{Mg(s)}) \\
&\quad + (I_1 + I_2) + \tfrac{1}{2}D_m(O-O) \\
&\quad + L_0(\text{MgO(s)}) \\
&= (601.7 + 147.7 + 736 + 1452 + 249 \\
&\quad - 3836) \\
&= -649.6\,\text{kJ}\,\text{mol}^{-1}
\end{aligned}
$$

$E(O) = 141$, so $E(O^-) = -649.6 - 141 = -790.6\,\text{kJ}\,\text{mol}^{-1}$.

This implies that the enthalpy change for the addition of an electron to the $O^-(g)$ anion is also endothermic:

$$
O^-(g) + e^-(g) = O^{2-}(g); \qquad \Delta H_m^{\ominus} = 790.6\,\text{kJ}\,\text{mol}^{-1}.
$$

22. First we use equation (1.17) to calculate a value for the lattice energy:

$$
L_0 = -\frac{1.214 \times 10^5 v Z_+ Z_-}{r_+ + r_-}\left(1 - \frac{34.5}{r_+ + r_-}\right).
$$

For NH_4Cl, $v = 2$, $Z_+ = 1$, $Z_- = 1$, $r_+ = 151\,\text{pm}$, $r_- = 167\,\text{pm}$, giving $L_0 = -681\,\text{kJ}\,\text{mol}^{-1}$. From Figure 1.48:

$$
\begin{aligned}
P(\text{NH}_3(g)) &= -\Delta H_f^{\ominus}(\text{NH}_4\text{Cl(s)}) + \Delta H_f^{\ominus}(\text{NH}_3(g)) \\
&\quad + \tfrac{1}{2}D_m(\text{H}-\text{H}) + I(\text{H}) + \tfrac{1}{2}D_m(\text{Cl}-\text{Cl}) - E(\text{Cl}) \\
&\quad + L_0(\text{NH}_4\text{Cl(s)}) \\
&= (314.4 - 46.0 + 218 + 1314 + 122 - 349 \\
&\quad - 681)\,\text{kJ}\,\text{mol}^{-1} \\
&= 892.4\,\text{kJ}\,\text{mol}^{-1}.
\end{aligned}
$$

The addition of a proton to the ammonia molecule is an exothermic process

$$
NH_3(g) + H^+(g) = NH_4^+(g); \qquad \Delta H_m^{\ominus} = -892.4\,\text{kJ}\,\text{mol}^{-1}.
$$

(The experimentally determined value is $871 \pm 15\,\text{kJ}\,\text{mol}^{-1}$, quite good agreement!)

23. The Born–Haber cycle for a chloride, MCl, is given in Figure 1.46 and leads to equation (1.2):

$$
\begin{aligned}
\Delta H_f^{\ominus}(\text{MCl(s)}) &= \Delta H_{\text{atm}}^{\ominus}(\text{M(s)}) + I_1(\text{M}) + \tfrac{1}{2}D_m(\text{Cl}-\text{Cl}) - E(\text{Cl}) \\
&\quad + L_0(\text{MCl(s)})
\end{aligned}
$$

giving a value for the lattice energy of NaCl of $L_0 = -786\,\text{kJ}\,\text{mol}^{-1}$. Assuming that this is also the value of L_0 for MgCl and AlCl, then we can calculate that

$$\Delta H_f^{\ominus}(\text{MgCl(s)}) = -129\,\text{kJ}\,\text{mol}^{-1}$$

and

$$\Delta H_f^{\ominus}(\text{AlCl(s)}) = -110\,\text{kJ}\,\text{mol}^{-1}.$$

$\Delta S_f^{\ominus}(\text{NaCl(s)}) = -90.6\,\text{J}\,\text{K}^{-1}\,\text{mol}^{-1}$, and using this value for MgCl and AlCl gives

$$\Delta G_f^{\ominus}(\text{MgCl(s)}) = -102\,\text{kJ}\,\text{mol}^{-1}$$

and

$$\Delta G_f^{\ominus}(\text{AlCl(s)}) = -83\,\text{kJ}\,\text{mol}^{-1}.$$

This may seem a surprising result as it tells us that both MgCl(s) and AlCl(s) are stable with respect to decomposition into their elements under standard conditions and yet we know that they do not exist! The answer is that there is in each case another thermodynamically favourable decomposition route, a **disproportionation**, that can take place:

$$2\text{MgCl(s)} = \text{Mg(s)} + \text{MgCl}_2(\text{s}); \qquad \Delta G_m^{\ominus} = -388\,\text{kJ}\,\text{mol}^{-1},$$
$$3\text{AlCl(s)} = 2\text{Al(s)} + \text{AlCl}_3(\text{s}); \qquad \Delta G_m^{\ominus} = -380\,\text{kJ}\,\text{mol}^{-1}.$$

<table>
<tr><td>2</td><td># Bonding in solids and electronic properties</td></tr>
</table>

2.1 INTRODUCTION

In the first chapter, we introduced the physical structure of solids – how their atoms are arranged in space. We now turn to a description of the bonding in solids – the electronic structure. Some solids consist of molecules bound together by very weak forces. We shall not be concerned with these because their properties are essentially those of the molecules. Nor shall we be much concerned with 'purely ionic' solids bound by electrostatic forces between ions as discussed in Chapter 1. The solids considered here are those in which all the atoms can be regarded as bound together. To illustrate how this bonding is reflected in the properties of the solids, we explore electronic properties of various types of solid.

Solids display a wide variety of interesting and useful electronic properties. Good electronic conductivity is one of the characteristic properties of metals; semiconductors are the foundation of the 'silicon revolution'. But why is tin a metal, silicon a semiconductor and diamond an insulator? Many solid state devices (transistors, photocells, light-emitting diodes (LEDs), solid state lasers, solar cells) are based on semiconductors containing carefully controlled amounts of impurity. How do these impurities affect the conductivity? These are some of the questions to be addressed, but first the basic bonding theory will be introduced.

2.2 BONDING IN SOLIDS – BAND MODEL

Traditionally bonding in metals has been approached through the idea of free electrons, a sort of electron gas. We discuss this model briefly in this chapter when we come to consider simple metals such as sodium and aluminium. However for most purposes, it is more fruitful to think of solids as a very large collection of atoms bonded together. For chemists this has the advantage that solids are not treated as totally different species from small molecules.

The theory most often used by chemists to describe the way electrons bind nuclei together to form molecules is the molecular orbital theory. This assumes that electrons have wave-like properties and are described by wave functions that result from the interaction of the electrons with all the nuclei in the molecule. The equation that is used to calculate the wave functions is the **Schrödinger equation**. Solving such an equation for a solid is something of a tall order however since exact solutions have not yet been found for small molecules and even a small crystal could well contain of the order of 10^{20} atoms. An approximation often used for smaller molecules is that the molecular wave functions can be formed by combining atomic wave functions. This **linear combination of atomic orbitals (LCAO)** approach can also be applied to solids.

We shall start by reminding the reader how to combine atomic orbitals for a very simple molecule, H_2. For H_2 we assume that the molecular orbitals are formed by combining $1s$ orbitals on each of the hydrogen atoms. These can combine in phase to give a bonding orbital or out of phase to give an anti-bonding orbital. The bonding orbital is lower in energy than the $1s$ and the anti-bonding orbital higher in energy. The amount by which the energy is lowered for a bonding orbital depends on the amount of overlap of the $1s$ orbitals on the two hydrogens. If the hydrogen nuclei are pulled further apart, for example, the overlap decreases and so the decrease in energy is less. (If the nuclei are pushed together, the overlap will increase but the electrostatic repulsion of the two nuclei becomes important and counteracts the effect of increased overlap.)

Suppose we form a chain of hydrogen atoms. For N hydrogen atoms, there will be N molecular orbitals. The lowest energy orbital will be that in which all the $1s$ orbitals combine in phase, and the highest energy orbital will be that in which the orbitals combine out of phase. In between are $(N - 2)$ molecular orbitals in which there is some in phase and some out of phase combination. Figure 2.1 shows a plot of the energy levels as the length of the chain increases. Note that as the number of atoms increases, the number of levels increases but the spread of energies seems to increase only slowly and is levelling off for long chains. Extrapolating to crystal length chains, one can see that there would be a very large number of levels within a comparatively small range of energies. A chain of hydrogen atoms is a very simple and artificial model; as an estimate of the energy separation of the levels, let us take a typical band in an average size metal crystal. A metal crystal might contain 10^{20} atoms and the range of energies be only 10^{-19} J. The average separation between levels would thus be only 10^{-39} J. The lowest energy levels in the hydrogen atom are separated by energies of the order of 10^{-18} J, showing that the

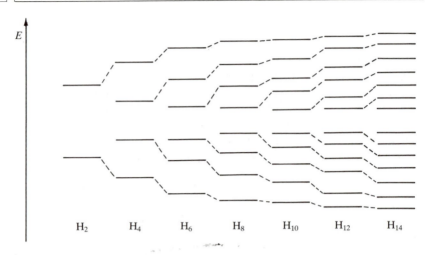

Figure 2.1 Orbital energies for a chain of N hydrogen atoms as N increases.

energy separation in a crystal is minute. The separation is in fact so small that we can think of the set of levels as a continuous range of energies. Such a continuous range of allowed energies is known as an **energy band**.

In place of a set of discrete levels then, we have an energy band. The hydrogen chain orbitals were made up from only one sort of atomic orbital, $1s$, and one energy band was formed. For most of the other atoms in the Periodic Table it is necessary to consider other atomic orbitals in addition to $1s$. Aluminium for example has the atomic configuration $1s^2 2s^2 2p^6 3s^2 3p^1$ and would be expected to form a $1s$ band, a $2s$ band, a $2p$ band, a $3s$ band and a $3p$ band. In fact the lower energy bands, those formed from the core orbitals $1s$, $2s$ and $2p$, are very narrow and for most purposes can be regarded as a set of localized atomic orbitals. This arises because these orbitals are concentrated very close to the nuclei and so there is little overlap between orbitals on neighbouring nuclei. In small molecules, the greater the overlap the greater the energy difference between bonding and anti-bonding orbitals. Likewise for continuous solids, the greater the overlap, the greater the spread of energies or **band width** of the resulting band. For aluminium, then, $1s$, $2s$ and $2p$ electrons can be taken to be core orbitals and only $3s$ and $3p$ bands considered.

Just as in small molecules, the available electrons are assigned to levels in the energy bands starting with the lowest. Each orbital can take two electrons of opposed spin. Thus if N atomic orbitals were combined to make the band orbitals, then $2N$ electrons are needed to fill the band. For example, the $3s$ band in a crystal of aluminium containing N atoms can take up to $2N$ electrons, whereas the $3p$ band can accommodate up to $6N$ electrons. As aluminium has only one $3p$ electron per atom, however, there would only be N electrons in the $3p$ band and only $N/2$ levels would be occupied. The highest occupied level at $0\,K$ is called the **Fermi level**. At temperatures above $0\,K$, some electrons just below the Fermi level will be excited to levels just above the Fermi level, but the number is small compared to the number of electrons in the band and so for most purposes, we make the approximation that electrons occupy all levels from the bottom of the band up to the Fermi level. Now we look at some examples of solids and see how we can apply the concept of energy bands to understanding some of their properties.

2.3 ELECTRONIC CONDUCTIVITY – SIMPLE METALS

The elements on the far left of the Periodic Table – Group 1 (sodium, potassium, etc.), Group 2 (magnesium, calcium, etc.) and aluminium – are often described as simple metals. The crystal structures of these metals are such that the atoms have high coordination numbers. For example, the Group 1 elements have body-centred cubic structures with each atom surrounded by eight others. This high coordination number increases the number of ways in which the atomic orbitals can overlap. The ns and np bands of the simple metals are very wide, due to the large amount of overlap, and, since the ns and np atomic orbitals are relatively close in energy, the two bands merge. This can be shown even for a small chain of lithium atoms (Figure 2.2). For the simple metals we then do not have an ns band and an np band but rather one continuous band which we shall label ns/np. For a crystal of N atoms this ns/np band contains $4N$ energy levels and can hold up to $8N$ electrons.

The simple metals have far fewer than $8N$ electrons available however; they have only N, $2N$ or $3N$. Thus the band is only partly full. In a partly full band, it requires very little energy to promote an electron to a higher energy level and because this promoted electron is delocalized, it can transport its energy to other positions in the crystal. This ease of promotion is crucial in explaining characteristic properties of metals such as high electronic and thermal conductivity which involve the ease of transport of energy through a solid. The high electrical conductivity of metals was however first explained in terms of a simpler theory than the band model, the free electron model. A fuller explanation of the electronic conductivity of simple metals will be left until after this latter model has been introduced.

2.3.1 Free electron theory

The free electron model regards a metal as a box in which electrons are free to roam unaffected by the atomic nuclei or by each other. The simple

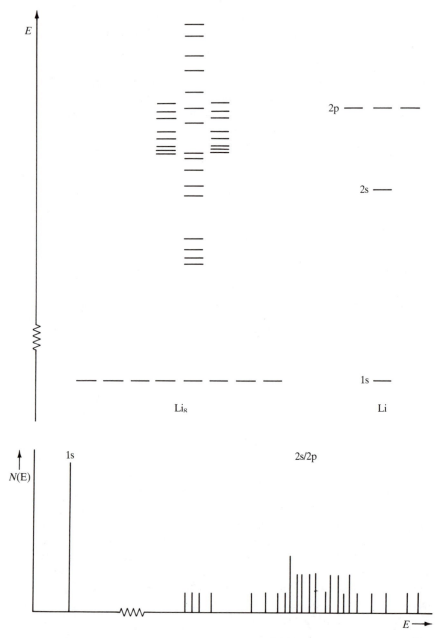

Figure 2.2 Orbital energies for a chain of eight lithium atoms.

metals provide the nearest approximation to this model as the electrons in the ns/np band spend most of their time between the nuclei and can keep well away from each other.

The model assumes that the nuclei stay fixed on their lattice sites surrounded by the inner or core electrons whilst the outer or valence electrons travel freely through the solid. If we ignore the cores then the quantum mechanical description of the outer electrons becomes very simple. Taking just one of these electrons the problem becomes the well-known one of the '**particle in a box**'. We start by considering an electron in a one-dimensional solid. The electron is confined to a line of length a (the length of the solid) which we shall call the x-axis. Since we are ignoring the cores, there is nothing for the electron to interact with and so it experiences zero potential within the solid. The Schrödinger equation for the electron is

$$-\hbar^2/2m_e \, d^2\psi/dx^2 = (E - V)\psi, \tag{2.1}$$

where $\hbar$ is Planck's constant divided by 2π, m_e is the mass of the electron, V is the electrical potential, ψ is the wave function of the electron and E is the energy of an electron with that wave function. When $V = 0$, the solutions to this equation are simple sine or cosine functions, and this can be verified by substituting $\psi = \sin\sqrt{2m_e E/\hbar^2}\, x$ into equation (2.1), as follows.

If $\psi = \sin\sqrt{2m_e E/\hbar^2}\, x$ then differentiating once gives

$d\psi/dx = \sqrt{2m_e E/\hbar^2} \cos\sqrt{2m_e E/\hbar^2}\, x \cdot$

Differentiating twice, gives $d^2\psi/dx^2 = -2m_e E/\hbar^2 \sin\sqrt{2m_e E/\hbar^2}\, x$

which is $d^2\psi/dx^2 = -2m_e E/\hbar^2\, \psi$ corresponding to (2.1) rearranged and with $V = 0$.

The electron is not allowed outside the box and to ensure this we put the potential to infinity outside the box. Since the electron cannot have infinite energy, the wave function must be zero outside the box and since it cannot be discontinuous, it must be zero at the boundaries of the box. If we take the sine wave solution, then this is zero at $x = 0$. To be zero at $x = a$, there must be a whole number of half waves in the box. Sine functions have a value of zero at angles of $n\pi$ radians where n is an integer and so $a\sqrt{2m_e E/\hbar^2} = n\pi$. The energy is thus quantized, $E = n^2 h^2/(8m_e a^2)$, with quantum number n. Since n can take all integral values this means there are an infinite number of energy levels with larger and larger gaps between each level. Most solids of course are three-dimensional (although we shall meet some later where conductivity is confined to one or two dimensions) and so we need to extend the free electron theory to three dimensions.

For three dimensions the metal can be taken as a rectangular box $a \times b \times c$. The appropriate wave function is now the product of three sine or cosine functions and the energy is given by

$$E = h^2(n_a^2/a^2 + n_b^2/b^2 + n_c^2/c^2)/8m_e. \tag{2.2}$$

Each set of quantum numbers n_a, n_b, n_c will give rise to an energy level. However in three dimensions there are many combinations of n_a, n_b and n_c that will give the same energy, whereas for the one-dimensional model there were only two levels of each energy (n and $-n$). For example, the following sets of numbers all give $(n_a^2/a^2 + n_b^2/b^2 + n_c^2/c^2) = 108$

n_a/a	n_b/b	n_c/c
6	6	6
2	2	10
2	10	2
10	2	2

and hence the same energy. The number of states with the same energy is known as the **degeneracy**. For small values of the quantum numbers, it is possible to write out all the combinations that will give rise to the same energy. If we are dealing with a crystal of say 10^{20} atoms it becomes difficult to work out all the combinations especially near the Fermi level. To estimate the degeneracy of any level in a band of this size we introduce a quantity called the wave vector. If we substitute k_x, k_y and k_z for $n_a\pi/a$, $n_b\pi/b$ and $n_c\pi/c$, then the energy becomes

$$E = (k_x^2 + k_y^2 + k_z^2)\hbar^2/2m_e. \tag{2.3}$$

k_x, k_y and k_z can be considered as the components of a vector, $\mathbf{k}$; the energy is proportional to the square of the length of this vector. $\mathbf{k}$ is called the **wave vector** and is related to the momentum of the electron wave as can be seen by comparing the classical expression $E = p^2/2m$, where p is the momentum, with the expression above. This gives the electron momentum as $\pm k\hbar$.

All the combinations of quantum numbers giving rise to one particular energy correspond to a wave vector of the same length $|\mathbf{k}|$. Thus all possible combinations leading to a given energy produce vectors whose ends lie on the surface of a sphere of radius $|\mathbf{k}|$. The total number of wave vectors with energies up to and including that with the given energy is given by the volume of the sphere, that is $4k^3\pi/3$ where $|\mathbf{k}|$ is written as k. The number of states with energies up to the given energy depend on n_a, etc., rather than k_x, etc., and so we have to multiply by abc/π^3. But how many states are there with the particular energy? Well it is actually more useful to define the number of states in a narrow range. Let us take a range of k values dk. The number of states up to and including those of wave vector length $k + dk$ is $4/3\pi^2 \ V(k + dk)^3$, where

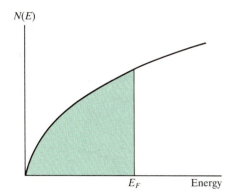

Figure 2.3 A density of states curve based on the free electron model. The levels occupied at 0 K are shaded. Note that later in the book, energy is plotted on the vertical axis. In this figure energy is plotted along the horizontal axis for comparison with the experimental results in Figure 2.4.

V $(= abc)$ is the volume of the crystal. So the number with values between k and $k + dk$ is $4/3\pi^2 \, V((k + dk)^3 - k^3)$ which when $(k + dk)^3$ is expanded gives a leading term $4/\pi^2 \, V k^2 dk$. This quantity is the **density of states**, $N(k) \, dk$. In terms of the more familiar energy, the density of states $N(E) \, dE$ is given by $(2m_e)^{3/2} V E^{1/2}/2\pi^2 \hbar^3 \, dE$. A plot of $N(E) \, dE$ against E is given in Figure 2.3.

The concepts of wave vectors and density of states are not confined to free electron theory. In the orbital theory, the energy levels within a band have varying degeneracies and so we can plot the number of levels with a particular range of energy against energy just as for the free electron model. Figure 2.2b was a density of states diagram for Li_8. The density of states diagram produced by the orbital theory will not, however, obey a simple formula like the one shown in Figure 2.3, but will vary from band to band. In general though, orbital density of states diagrams show fewer energy levels, that is a lower density of states, at the top and bottom of the band and more, that is a higher density of states, in the middle.

Experimentally the density of states can be determined by X-ray emission spectroscopy. A beam of electrons or high energy X-rays hitting a metal can remove core electrons. In sodium for example the $2s$ or $2p$ electrons might be removed. The core energy levels are essentially atomic levels and so electrons have been removed from a discrete well-defined energy level. Electrons from the conduction band can now jump down to this energy level, emitting X-rays as they do so. The X-ray energy will depend on the level of the conduction band from which the electron has come. A scan across the emitted X-rays will correspond to a scan across the filled levels of the conduction band. The intensity of the radiation

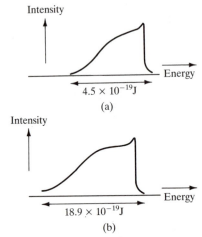

Figure 2.4 X-ray emission spectra obtained from (a) sodium metal and (b) aluminium metal when conduction electrons fall into the $2p$ level. The slight tail at the high energy end is due to thermal excitation of electrons close to the Fermi level.

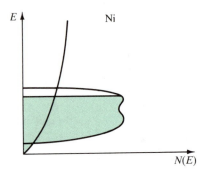

Figure 2.5 The band structure of nickel. Note the shape of the $3d$.

emitted will depend on the number of electrons with that particular energy, that is the intensity depends on the density of states of the conduction band. Figure 2.4 shows some X-ray emission spectra for sodium and aluminium and one can see that the shape of these curves resembles approximately the occupied part of Figure 2.3, so that the free electron model appears to describe these bands quite well.

These bands are, however, less than half full and if we looked at the unoccupied part or at a metal with a band that was more than half full, we should find that instead of continuing to increase as in Figure 2.3, the density of states would start to decrease with energy and reach zero at the top of the band. This is shown, for example, in the nearly full $3d$ band of Ni given in Figure 2.5.

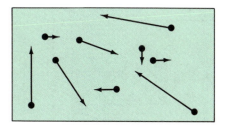

Figure 2.6 Electrons in a metal in the absence of an electric field. They move in all directions but overall there is no net motion in any direction.

2.3.2 Electronic conductivity

To explain electronic conductivity, it is necessary to turn to the wave vector. This concept was derived for a free electron in a box, but it can also be used in the orbital method. The combinations of atomic orbitals forming crystal orbitals form periodic patterns like the sine waves that were needed to fit into the box of the free electron model. For example, along a particular direction the phase of the atomic orbital might change sign every other atom or every third or fourth atom.

Such combinations can be represented by a term in the coefficient $e^{i\mathbf{k}\cdot\mathbf{r}}$, where $\mathbf{r}$ is the position vector of a nucleus in the lattice and $\mathbf{k}$, as in free electron theory, represents the momentum of the electron. The LCAO expression for the orbitals

$$\Psi(i) = \sum_n c_{ni}\phi_n$$

(where the sum is over all n atoms and $\Psi(i)$ is the ith orbital and c_{ni} is the coefficient of the orbital on the nth atom for the ith molecular orbital) is replaced by

$$\Psi(\mathbf{k}) = \sum_n e^{i\mathbf{k}\cdot\mathbf{r}_n} a_{n\mathbf{k}}\phi_n,$$

where $\Psi(\mathbf{k})$ is the wave function with wave vector $\mathbf{k}$ and c_{ni} is replaced by $e^{i\mathbf{k}\cdot\mathbf{r}_n} a_{n\mathbf{k}}$.

Whichever model of bonding we use, $\mathbf{k}$ represents a quantized electron momentum, and for explaining electronic conductivity it is important to note that it is a vector with direction as well as magnitude. Thus there may be many different orbitals with the same value of k and hence the same value of the energy, but with different components, k_x, k_y, k_z, giving a different direction to the momentum. In the absence of an electric field, all directions of the wave vector $\mathbf{k}$ are equally likely and so there are equal numbers of electrons moving in all directions (Figure 2.6).

If we now connect our metal to the terminals of a battery producing an electric field, then an electron travelling in the direction of the field will

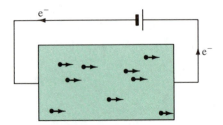

Figure 2.7 The sample of Figure 2.5 in a constant electric field, established by placing the rod between the terminals of a battery. The electrons continue to move in all directions but now their velocities are modified so that each also has a net movement or drift velocity in the left to right direction.

be accelerated and the energies of those levels with a net momentum in this direction will be lowered. Electrons moving the other way will have their energies raised and so some of these electrons will drop down into levels of lower energy corresponding to momentum in the opposite direction. There will thus be more electrons moving one way than the other. This is an electric current. The net velocity in an electric field is shown in Figure 2.7.

The electronic conductivity, σ, is given by the expression

$$\sigma = nZe\mu,$$

where n is the number of charge carriers per unit volume, Ze is their charge (in the case of an electron this is simply e the electronic charge) and μ is a measure of the velocity in the electric field.

The electrons in the levels of higher energy can only move into levels of lower energy if these are empty, and there will only be empty levels if the band is partly full. Thus *only solids with partly occupied energy bands are good electronic conductors*. Note also that only electrons near the Fermi level can be promoted and hence only these carry the current. Thus the factor n in the expression above, and hence the conductivity, will depend on the density of states near the Fermi level in the partly filled band.

Although the possession of a partly full band provides an explanation of electronic conductivity, it does not explain all the features of this phenomenon. For example, it does not account for the finite resistance of metals. The current, i, flowing through a metal for a given applied electric field, V, is given by **Ohm's law**, $V = iR$, where R is the resistance. It is characteristic of a metal that R increases with increasing temperature (or putting it another way, σ decreases with increasing temperature), i.e. for a given field, the current decreases as the temperature is raised. There is nothing in our theory yet that will impede the flow of electrons. To account for electrical resistance, it is necessary to introduce the ionic cores. If these were arranged periodically on the lattice sites of a perfect crystal and did not move, then they would not interrupt the flow of electrons. Most crystals contain some imperfections however and these can scatter the electrons. If the component of the electron's momentum in the field direction is reduced then the current will drop. In addition, even in perfect crystals the ionic cores will be vibrating. There are a set of crystal vibrations in which the ionic cores vibrate together. These vibrations are called **phonons**. An example of a crystal vibrational mode would be one in which each ionic core would be moving out of phase with its neighbours along one axis. (If they moved in phase, the whole crystal would move.) Like vibrations in small molecules, each crystal vibration has its set of quantized vibrational levels. The conduction electrons moving through the crystal are scattered by the vibrating ionic cores and lose some energy to the phonons. As well as reducing the electron flow, this mechanism increases the crystal's vibrational energy. The effect then is to convert electrical energy to heat. This ohmic heating effect is put to good use in, for example, the heating elements of kettles.

2.4 SEMICONDUCTORS – SILICON AND GERMANIUM

Carbon (as diamond polymorph), silicon and germanium, instead of forming structures of high coordination like the simple metals, form structures in which the atoms are tetrahedrally coordinated. With these structures, the ns/np bands still overlap but the ns/np band splits into two. Each of the two bands contains $2N$ orbitals and so can accommodate up to $4N$ electrons. Think of the two bands as an analogy of bonding and anti-bonding; the tetrahedral symmetry not giving rise to any non-bonding orbitals. Carbon, silicon and germanium have electronic configurations $ns^2 np^2$ and so they have available $4N$ electrons, just the right number to fill the lower band. This lower band is known as the **valence band**, the electrons in this band essentially bonding the atoms in the solid together.

Why do these elements adopt the tetrahedral structure rather than one of the higher coordination structures? Well, if these elements adopted a structure like that of the simple metals, then the ns/np band would be half full. The electrons in the highest occupied levels would be virtually non-bonding. In the tetrahedral structure however all $4N$ electrons would be in bonding levels. This is illustrated in Figure 2.8. Elements with few valence electrons will thus be expected to adopt high coordination structures and be metallic. Those with larger numbers (four or more) will be expected to adopt lower coordination structures in which the ns/np band is split and only the lower bonding band is occupied.

Tin (in its most stable room temperature form) and lead, although in

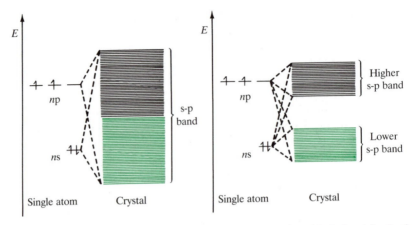

Figure 2.8 Energy bands formed from ns and np atomic orbitals for (a) a body-centred cubic crystal and (b) a crystal of diamond structure, showing filled levels for $4N$ electrons.

the same group as silicon and germanium, are metals. For these elements the atomic $s{-}p$ energy separation is greater and the overlap of s and p orbitals is much less than in silicon and germanium. For tin the tetrahedral structure would have two s/p bands but the band gap is almost zero. Below 291 K tin undergoes a transition to the diamond structure, but above this temperature it is more stable for tin to adopt a higher coordination structure. The advantage of having no non-bonding levels in the diamond structure is reduced by the small band gap. In lead, the diamond structure would give rise to an s band and a p band rather than ns/np bands since the overlap of s and p orbitals is even smaller. It is more favourable for lead to adopt a cubic close-packed structure and, because lead has only $2N$ electrons to go in the p band, it is metallic. This is an example of the **inert pair effect**, in which the s electrons act as core electrons, on the chemistry of lead. Another example is the formation of divalent ionic compounds containing Pb^{2+} rather than tetravalent covalent compounds like those of silicon and germanium.

Silicon and germanium have a completely full valence band and would be expected to be insulators. They do, however, belong to a class of materials known as **semiconductors**.

The conductivity of metallic conductors decreases with temperature. As the temperature is raised, the phonons gain energy; the lattice vibrations have larger amplitudes. The displacement of the ionic cores from their lattice sites is thus greater and the electrons are scattered more, reducing the net current, by reducing the mobility, μ, of the electrons.

The conductivity of **intrinsic semiconductors** such as silicon or germanium, however, *increases* with temperature. In these solids, conduction can only occur if electrons are promoted to the higher s/p band, the **conduction band**, because only then will there be a partially full band. The current in semiconductors will depend on the number of electrons free to transport charge. This is the number of electrons promoted to the conduction band plus the number in the valence band that have been freed to move by this promotion. As the temperature increases, the number of electrons promoted increases and so the current increases. At any one temperature more electrons will be promoted for a solid with a small band gap than for one with a large band gap so that the solid with the smaller band gap will be a better conductor. The number of electrons promoted varies with temperature in an exponential manner so that, for example, in germanium which has a band gap 10 times smaller than diamond, of the order of 10^{91} more electrons are promoted than in diamond. Germanium has an electrical resistivity of $0.46\,\Omega\,m$ at room temperature compared to that of a typical insulator of around $10^{12}\,\Omega\,m$.

2.4.1 Photoconductivity

Electrons can also be promoted by forms of energy other than heat energy, e.g. light. If the photon energy ($h\nu$) of light shining on a semiconductor is greater than the energy of the band gap, then valence band electrons will be promoted to the conduction band and conductivity will increase. Promotion of electrons by light is illustrated in Figure 2.9.

Semiconductors with band gap energies corresponding to photons of visible light are **photoconductors**, being essentially non-conducting in the dark but conducting electricity in the light. One use of such photoconductors is in **electrophotography**. In the xerographic process, there is a positively charged plate covered with a film of semiconductor. (In practice this semiconductor is not silicon but a solid with a more suitable energy band gap such as selenium or the compound As_2Se_3.) Light reflected from the white parts of the page to be copied hits the semiconductor film. The parts of the film receiving the light become conducting; an electron is promoted to the conduction band. This electron then cancels the positive charge on the film, the positive hole in the valence band being removed by an electron from the metal backing plate entering the valence band. Now the parts of the film which received light from the original are no longer charged, but the parts underneath the black lines are still positively charged. Tiny negatively charged plastic capsules of ink (toner) are then spread on to the semiconductor film but only stick to the charged bits of the film. A piece of positively charged white paper removes the toner from the semiconductor film and hence acquires an image of the black parts of the original. Finally the paper is heated to melt the plastic coating and fix the ink.

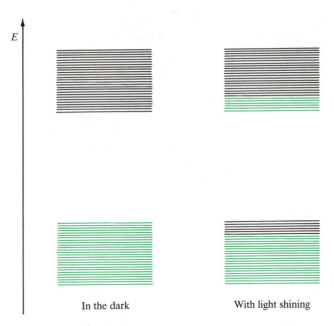

In the dark   With light shining

Figure 2.9 Promotion of electrons from the valence to the conduction band by light.

2.5 DOPED SEMICONDUCTORS

The conductivity of semiconductors can also be increased by introducing a very low concentration of impurity. Such semiconductors are known as doped or extrinsic semiconductors. Consider a crystal of silicon containing boron as an impurity. Boron has one fewer valence electron than silicon and so for every silicon replaced by boron there is an electron missing from the valence band (Figure 2.10), i.e. there are positive holes in the valence band and these enable electrons near the top of the band to conduct electricity. So the doped solid will be a better conductor than pure silicon. A semiconductor like this doped with an element with fewer valence electrons than the bulk of the material is called a **p-type semiconductor** because its conductivity is related to the number of positive holes produced by the impurity.

Suppose instead of boron the silicon was doped with an element with more valence electrons than silicon, e.g. phosphorus. The doping atoms form a set of energy levels that lie in the band gap between the valence and conduction bands. Since the atoms have more valence electrons than silicon, these energy levels are filled. There are therefore electrons present close to the bottom of the conduction band and easily promoted

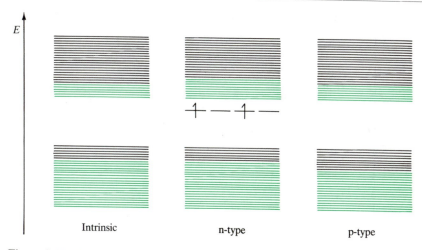

Figure 2.10 Intrinsic, *n*-type and *p*-type semiconductors showing negative charge carriers (electrons in the conduction band) and positive holes.

into the band. This time the conductivity increases because of extra electrons entering the conduction band. Such semiconductors are called *n*-**type**; *n* for negative charge carriers or electrons. Figure 2.10 shows schematically energy bands in intrinsic, *p*-type and *n*-type semiconductors.

n- and *p*-type semiconductors in various combinations make up many electronic devices. Rectifiers, field-effect transistors, photo-voltaic cells and light-emitting diodes are made from a crystal doped to be partly *n*- and partly *p*-type. Early transistors were made from *n*–*p*–*n* or *p*–*n*–*p* sandwiches. Other devices use mixtures of *p*- and/or *n*-type semiconductor with metal and insulator layers to produce the required properties. In the next section we consider what happens at the junction of two differently doped regions, and how such a junction is used as the basis of solar cells.

2.5.1 The *p*–*n* junction – photovoltaic (solar) cells

In the region of the crystal (or glass) where *n*- and *p*-type meet, there is a discontinuity in electron concentration. Although both *n*- and *p*-type are electrically neutral, the *n*-type has a greater concentration of electrons than the *p*-type. In order to try and equalize the electron concentrations electrons drift from *n*- to *p*-type. However, this produces a positive electric charge on the *n*-type and a negative electric charge on the *p*-type. The electric field thus set up encourages electrons to drift back to the *n*-type. Eventually a state is reached in which the two forces

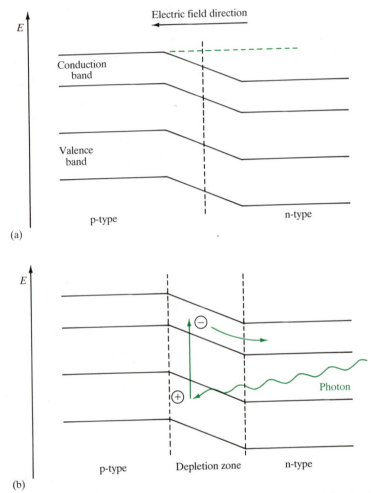

Figure 2.11 (a) Bending of energy levels across a *p–n* junction and (b) the effect of light on a *p–n* junction.

are balanced and the electron concentration varies smoothly across the junction as in Figure 2.11.

This is the situation in the dark. The use of the *p–n* junction as a solar cell depends on its response to light. If the wavelength is short enough for the photon energy to be greater than the band gap energy then electrons will be promoted across the band gap to the conduction band. The promoted electron will be attracted to the positively charged *n*-type region. The gap in the valence band will move to the *p*-type region because an electron from this side would be attracted across to the *n*-type region to fill any hole there. The promoted electron and the hole are now separated in space and so the conduction electron cannot simply emit light and return to the valence band. It is free, however, to travel through the *n*-type to an external circuit. The electric current thus produced can be used to do work; the illuminated *p–n* junction is acting as a battery. Such cells using sunlight as the source of illumination have found widespread application, although their efficiency is not high. They can be used for pocket calculators, for lighting homes and shops in remote areas, and in desert areas even for large-scale power generation. They are attractive because sunlight is a clean, cheap source of power. Semiconductors for these devices need a band gap in or slightly below the range of energies of visible light photons ($2.4 - 5.0 \times 10^{-19}$ J). Silicon has a band gap of 1.9×10^{-19} J and so can use all wavelengths of visible light.

2.6 BANDS IN COMPOUNDS – GALLIUM ARSENIDE

Gallium arsenide (GaAs) is a rival to silicon in some semiconductor applications, including solar cells, and is also used for light-emitting diodes and in a solid-state laser (Chapter 6). It has a diamond-type structure and is similar to silicon except that it is composed of two kinds of atoms. The valence orbitals in Ga and As are the $4s$ and $4p$ and these will form two bands, each containing $4N$ electrons as in silicon. Because of the different $4s$ and $4p$ atomic orbital energies in Ga and As, however, the lower band will have a greater contribution from As and the conduction band will have a higher contribution from Ga. Thus GaAs can be thought of as having partial ionic character because there is a partial transfer of electrons from Ga to As. The valence band has more As than Ga character and so all the valence electrons end up in orbitals in which the possibility of being near an As nucleus is greater than that of being near a Ga nucleus. The band energy diagram for GaAs is shown in Figure 2.12. GaAs is an example of a class of semiconductors known as III/V semiconductors in which an element with one more valence electron than the silicon group is combined with an element with one less valence electron. Many of these compounds are semiconductors, e.g. GaSb, InP, InAs and InSb. Moving further along the Periodic Table there are II/VI semiconductors such as CdTe and ZnS. Towards the top of the Periodic Table and further out towards the edges, e.g. AlN and AgCl, the solids tend to adopt different structures and become more ionic. For the semi-conducting solids, the band gap decreases down a group, e.g. GaP > GaAs > GaSb and AlAs > GaAs > InAs.

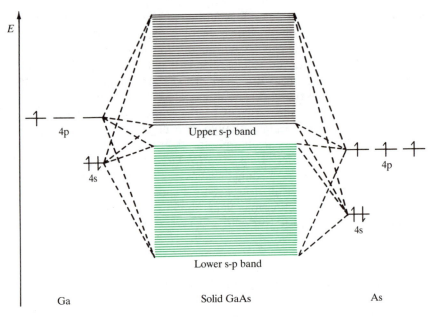

Figure 2.12 Orbital energy level diagram for gallium arsenide.

2.6.1 Semiconductor–liquid junction cells

n-Doped GaAs has been used in an interesting variant of the photovoltaic cell; the semiconductor–liquid junction cell. In these cells, the electric field is produced not where the *n*- and *p*-type semiconductors meet but where the *n*-type meets a liquid electrolyte. The cell consists of an electrolyte solution into which dip two electrodes, the *n*-type semiconductor and a metal electrode. When light shines on the *n*-GaAs, an electron is promoted to the conduction band, leaving a hole in the valence band. The conduction electron drifts towards the bulk of the semiconductor but the hole drifts towards the electrolyte. In the electrolyte solution is an ion that can exist in oxidized and reduced forms, e.g. Fe^{2+}/Fe^{3+}. The hole in the valence band is filled by an electron from the reduced form of the ion, the ion being oxidized. Reduction of the ion back to its reduced form takes place at the metal electrode and the circuit is completed by electrons travelling through the external circuit from the semiconductor to the metal. The electrons in the external circuit can do work and so the net result is that light is converted to electrical energy. Examples of electrolyte solutions that have been used include aqueous Se^{2-}/Se_n^{2-} and ferrocene/ferrocinium in acetonitrile.

2.7 BANDS IN *d*-BLOCK COMPOUNDS – TRANSITION METAL MONOXIDES

Monoxides (MO) with structures based on sodium chloride* are formed by the first row transition elements Ti, V, Mn, Fe, Co and Ni. TiO and VO are metallic conductors, and the others are semiconductors. The O $2p$ orbitals form a filled valence band. The $4s$ orbitals on the metal form another band. What of the $3d$ orbitals?

In the sodium chloride structure, the symmetry enables three of the five *d* orbitals on different metal atoms to overlap. Because the atoms are not nearest neighbours, the overlap is not as large as in pure metals and the bands are thus narrow. The other two *d* orbitals overlap with orbitals on the adjacent oxygens. There are thus two narrow $3d$ bands, the lower one, labelled t_{2g}, can take up to $6N$ electrons and the upper one, labelled e_g, up to $4N$ electrons. Divalent titanium has two *d* electrons and so there are $2N$ electrons to fill the $3N$ levels of the lower band. As in the case of pure metals, a partly filled band leads to metallic conductivity. However, on this basis one would expect MnO, CoO and NiO, which also have partly filled bands, to be metallic. These oxides are semiconductors though.

Going across the first transition series, there is a contraction in the size of the $3d$ orbitals. $3d$ orbital overlap therefore decreases and the $3d$ band narrows. In a wide band such as the s/p bands of the alkali metals, the electrons are essentially free to move through the crystal keeping away from the nuclei and from each other. In a narrow band by contrast, the electrons are more tightly bound to the nuclei. Interelectron repulsion becomes important; in particular repulsion between electrons on the same atom. Consider an electron in a partly filled band moving from one nucleus to another. In the alkali metal, the electron would already be in the sphere of influence of surrounding nuclei and would not be greatly repelled by electrons on these nuclei. The $3d$ electron moving from one nucleus to another adds an extra electron near to the second nucleus which already has $3d$ electron density near it. Thus electron repulsion on the nucleus is increased. For narrow bands therefore we have to balance gains in energy on band formation against electron repulsion. For MnO, FeO, CoO and NiO electron repulsion wins and it becomes more favourable for the $3d$ electrons to remain in localized orbitals than to be delocalized.

The band gap between the oxygen $2p$ band and the metal $4s$ band is sufficiently wide that the pure oxides would be considered insulators.

---

* As explained in Chapter 3 TiO and VO have structures based on sodium chloride but with $\frac{1}{6}$ of each element missing in an ordered manner. The other oxides when stoichiometric adopt the sodium chloride structure.

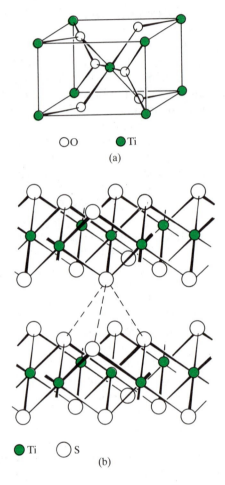

Figure 2.13   Crystal structures of (a) TiO$_2$ and (b) TiS$_2$.

However, they are almost invariably found to be non-stoichiometric, i.e. their formulae are not exactly MO, and this leads to semiconducting properties which will be discussed in Chapter 3.

2.7.1 Titanium dioxide and disulphide

The monoxides are not unique in displaying a variation of properties across the transition series. The dioxides from another series and CrO$_2$ will be discussed later because of its magnetic properties. Several classes of mixed oxides also exhibit a range of electronic properties (e.g. the

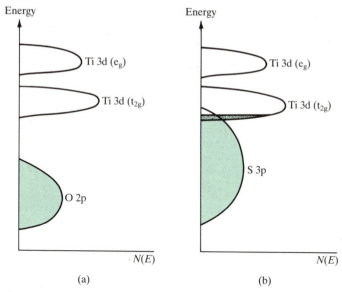

Figure 2.14 Energy bands for (a) TiO_2 and (b) TiS_2.

perovskites $LaTiO_3$, $SrVO_3$ and $LaNiO_3$ are metallic conductors, $LaRhO_3$ is a semiconductor and $LaMnO_3$ is an insulator). Sulphides also show progression from metal to insulator. In general, compounds with broader *d* bands will be metallic. Broad bands will tend to occur for elements at the beginning of the transition series and for second and third row metals, e.g. NbO and WO_2. Metallic behaviour is also more common amongst lower oxidation state compounds and with less electronegative anions.

The influence of the anion is illustrated by comparing TiO_2 with TiS_2. TiO_2 has the rutile structure (Figure 2.13) and is an insulator with a band gap energy just outside the range of photon energies for visible light. *n*-Doped TiO_2 finds use as a semiconductor, for example, in liquid junction photovoltaic cells. TiS_2 has a different, layer, structure (CdI_2 type) and is a semiconductor.

The band diagram for TiO_2 is shown in Figure 2.14. The O 2*p* band is full and the 3*d* (t_{2g}) band empty. The 3*d* energy levels are delocalized and so the lower 3*d* band forms the conduction band. In TiS_2, the overlap of S and Ti orbitals is greater than that of O and Ti orbitals in TiO_2 and the greater width of the S 3*p* band combined with the smaller S 3*p*/Ti 3*d* energy gap means that the 3*p* band is very close to the 3*d* band. The bands either just overlap or are separated by a very small band gap. Since the valence band electrons are thus very easily promoted, the conductivity is much higher than that of a wide band gap semiconductor. So close to a band edge, however, there are relatively few orbitals and hence relatively few electrons available for promotion. Consequently the conductivity is not as high as that of a typical metal.

FURTHER READING

Moore, W.J. (1967) *Seven Solid States*, Chapters II and III, W.A. Benjamin Inc.
  Although rather dated, this is an unusual and generally very readable book. Much of it is at introductory level and the style is attractive. In the Silicon chapter, the preparative methods need updating, but the description and basic theory of semiconductors are still very relevant.
McWeeny, R. (1979) *Coulson's Valence*, Oxford University Press, Oxford.
  This well known and standard textbook on quantum chemistry contains a good chapter on band theory from a chemists point of view.
Cox, P.A. (1987) *Electronic Structure and Chemistry of Solids*, Chapters 1, 4 and 7, Oxford University Press, Oxford.
  A good book on solid state theory for chemists at a higher level.
West, A.R. (1988) *Basic Solid State Chemistry*, Chapters 2 and 7, John Wiley, New York.
  This book is a more compact version of the author's Solid State chemistry and its applications. Chapter 2 is a survey of types of bonding in solids, but with little detail. Chapter 7 deals with electrical properties mainly from a phenomenological and applications point of view and covers material relevant to chapters 3 and 7 as well as to this chapter.
Duffy, J.A. (1990) *Bonding, Energy Levels and Bands in Inorganic Solids*, Chapters 4 and 7, Longman, London.
  This book is at a fairly advanced level, but is written from a chemists point of view. Although containing material relevant to this chapter, it is at its most original and useful in its approach to optical properties.
Rosenberg, H.M. (1989) *The Solid State*, Chapters 7–10, Oxford University Press, Oxford.
  An elementary solid state physics book. It is one of the most accessible of the solid state texts written for physicists.
Kittel, C. (1986) *Introduction to Solid State Physics*, John Wiley, New York.
  A standard text for physics courses.

QUESTIONS

1. In the free electron model, the electron energy is entirely kinetic. Using the formula $E = \frac{1}{2}mv^2$, calculate the velocity of electrons at the Fermi level in sodium metal (Figure 2.4). The mass of an electron is 9.11×10^{-31} kg, and $1\,\text{eV} = 1.602 \times 10^{-19}$ J.

2. The density of magnesium metal is $1740\,\text{kg m}^{-3}$. A typical crystal has a volume of $10^{-12}\,\text{m}^3$ (corresponding to a cube of side 0.1 cm). How many atoms would such a crystal contain? The relative atomic mass of sodium is 23.

3. An estimate of the total number of occupied states can be obtained by integrating the density of states from 0 to the Fermi level.

$$N = \int_0^{E_F} N(E)\,dE = \int_0^{E_F} E^{1/2}(2m_e)^{3/2}V/(2\pi^2\hbar^3)\,dE$$

$$= (2m_e E_F)^{3/2} V/3\pi^2\hbar^3.$$

Calculate the total number of occupied states for a sodium crystal of volume (a) $10^{-12}\,m^3$, (b) $10^{-6}\,m^3$ and (c) $10^{-29}\,m^3$ (approximately atomic size). Compare your results with the number of electrons available and comment on the different answers to (a), (b) and (c). $E_F = 2.8$ for sodium.

4. The energy associated with one photon of visible light ranges from 2.4 to $5.0 \times 10^{-19}\,J$. The band gap in selenium is $2.9 \times 10^{-19}\,J$. Explain why selenium is a good material to use as a photoconductor in applications such as photocopiers.

5. The band gaps of several semiconductors and insulators are given below. Which substances would be photoconductors over the entire range of visible wavelengths?

Substance	Si	Ge	CdS
Band gap $(10^{-19}\,J)$	1.9	1.3	3.8

6. Which of the following doped semiconductors will be p-type and which will be n-type? (a) Arsenic in germanium, (b) germanium in silicon, (c) indium in germanium, (d) boron in indium antimonide (InSb).

7. Would you expect carborundum (SiC) to adopt a diamond structure or one of higher coordination? Explain why.

ANSWERS

1. For the Fermi level in sodium $E = 2.8\,eV$ and the mass of an electron is $9.11 \times 10^{-31}\,kg$. With $1\,eV = 1.602 \times 10^{-19}\,J$, this gives:

$$2.8 \times 1.602 \times 10^{-19} = \tfrac{1}{2} \times 9.11 \times 10^{-31} \times v^2$$
$$v = (2 \times 2.8 \times 1.602 \times 10^{-19}/9.11 \times 10^{-31})^{\tfrac{1}{2}}$$
$$= 9.9 \times 10^5\,ms^{-1}.$$

2. $10^{-12}\,m^3$ of metal contains $970 \times 10^{-12}\,kg$ of sodium, but one atom of sodium weighs $23/6.022 \times 10^{23}\,g = 23 \times 10^{-3}/6.022 \times 10^{23}\,kg$. So there are:

$970 \times 10^{-12} \times 6.022 \times 10^{23}/23 \times 10^{-3}$ atoms of sodium

$= 2.5 \times 10^{16}$ atoms.

3. (a) $N = 10^{-12} \times (2 \times 9.11 \times 10^{-31} \times 2.8 \times 1.602 \times 10^{-19})^{3/2}/3 \times \pi^2$
$\times (1.055 \times 10^{-34})^3 = 2.125 \times 10^{16}$.

(b) 2.125×10^{22}.

(c) 0.2.

Each level can take two electrons and a crystal of N atoms of sodium has N electrons to fill the band. As one can see, the agreement between the number of filled levels predicted by this very simple theory and the number needed to accommodate the available electrons is very good. Note also that this question illustrates how the energy level spacing increases as the electrons are confined to a smaller and smaller volume.

4. The band gap lies in the energy range for visible photons and so photons of visible light can promote electrons from the valence band to the conduction band. Electrons in both bands can then conduct electricity.

5. Si, Ge.

6. (a) n-type, (b) neither, (c) p-type, (d) p-type.

7. Carborundum, like silicon and germanium, has $4N$ valence electrons for a crystal of N atoms. The tetrahedral diamond structure will be favoured because all $4N$ electrons will then be in bonding orbitals and the energy is lower than in the higher coordination structure.

Defects and non-stoichiometry

3.1 INTRODUCTION

In a perfect crystal all atoms would be on their correct lattice positions in the structure. This situation can only exist at the absolute zero of temperature, 0 K. Above 0 K, **defects** occur in the structure. These defects may be **extended defects** such as **dislocations**. The strength of a material depends very much on the presence (or absence) of extended defects such as dislocations and **grain boundaries** but discussion of this type of phenomena lies very much in the realm of materials science and will not be discussed in this book. Defects can also occur at isolated atomic positions; these are known as **point defects**, and can be due to the presence of a foreign atom at a particular site or to a vacancy where normally one would expect an atom. Point defects can have significant effects on the chemical and physical properties of the solid. The beautiful colours of many gemstones are due to impurity atoms in the crystal structure. Ionic solids are able to conduct electricity by a mechanism which is due to the presence of vacant ion sites within the lattice (this is in contrast to the electronic conductivity explored in Chapter 2).

3.2 DEFECTS AND THEIR CONCENTRATION

Defects fall into two main categories: **intrinsic defects** which are integral to the crystal in question – they do not change the overall composition and because of this are also known as **stoichiometric defects**; and **extrinsic defects** which are created when a foreign atom is inserted into the lattice.

3.2.1 Intrinsic defects

Intrinsic defects fall into two categories: **Schottky defects**, which consist of vacancies in the lattice and **Frenkel defects** where a vacancy is created by an atom or ion moving into an interstitial position.

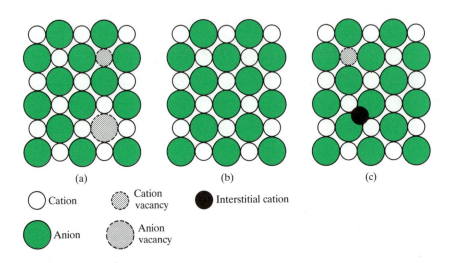

Figure 3.1 Schematic illustration of intrinsic point defects in a crystal of composition MX: (a) Schottky pair; (b) perfect crystal; (c) Frenkel pair.

For a 1:1 solid MX, a Schottky defect consists of a *pair* of vacant sites, a cation vacancy and an anion vacancy. This is shown in Figure 3.1 for an alkali-halide-type structure: the number of cation vacancies and anion vacancies have to be equal in order to preserve electrical neutrality. A Schottky defect for an MX_2-type structure will consist of the vacancy caused by the M^{2+} ion together with *two* X^- anion vacancies, thereby balancing the electrical charges. Schottky defects are more common in 1:1 stoichiometry and examples of crystals that contain them include rock salt (NaCl), wurtzite (ZnS) and CsCl.

A Frenkel defect usually only occurs on one **sublattice** of a crystal, and consists of an atom or ion moving into an interstitial position thereby creating a vacancy. This is illustrated in Figure 3.1c for an alkali-halide-type structure such as NaCl, where one cation is shown as having moved out of the lattice and into an interstitial site. This type of behaviour is seen, for instance, in AgCl where we observe such a **cation Frenkel defect** when Ag^+ ions move from their octahedral coordination sites into tetrahedral coordination (Figure 3.2). The formation of this type of defect is important in the photographic process when they are formed in the light-sensitive AgBr used in the photographic emulsions.

It is less common to observe an **anion Frenkel defect** when an anion moves into an interstitial site. This is because anions are commonly larger

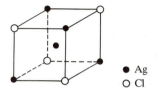

Figure 3.2 The tetrahedral coordination of an interstitial Ag^+ ion in AgCl.

than the cations in the structure and so it is more difficult for them to enter a crowded low-coordination interstitial site.

An important exception to this generalization lies in the formation of anion Frenkel defects in compounds with the fluorite structure, such as CaF_2 (other compounds adopting this structure are strontium and lead fluorides (SrF_2, PbF_2) and thorium, uranium and zirconium oxides (ThO_2, UO_2, ZrO_2) which we meet again later in this chapter). One reason for this is that the anions have a lower electrical charge than the cations and so do not find it as difficult to move nearer each other. The other reason lies in the nature of the fluorite structure (Figure 3.3). Recall that we can think of it as based on a ccp array of Ca^{2+} ions with all the tetrahedral holes occupied by the F^- ions. This of course leaves all of the larger octahedral holes unoccupied, giving a very open structure. This is shown clearly if we redraw the structure as in Figure 3.3c based on a simple cubic array of F^- ions. The unit cell now consists of eight small **octants** with the Ca^{2+} ions occupying every other octant. The two different views are completely equivalent, but the cell shown in Figure 3.3c shows the possible interstitial sites very clearly.

3.2.2 The concentration of defects

Energy is required to form a defect: this means that the formation of defects is always an **endothermic process**. It may seem surprising then that defects exist in crystals at all, and yet they do even at low temperatures, albeit in very small concentrations. The reason for this is that the formation of defects produces a commensurate gain in entropy. The enthalpy of formation of the defects is thus balanced by the gain in entropy such that, at equilibrium, the overall change in free energy of the crystal due to the defect formation is zero according to the equation:

$$\Delta G = \Delta H - T\Delta S.$$

The interesting point is that thermodynamically we do not expect a crystalline solid to be perfect, contrary, perhaps to our 'commonsense' expectation of symmetry and order! At any particular temperature there will be an equilibrium population of defects in the crystal.

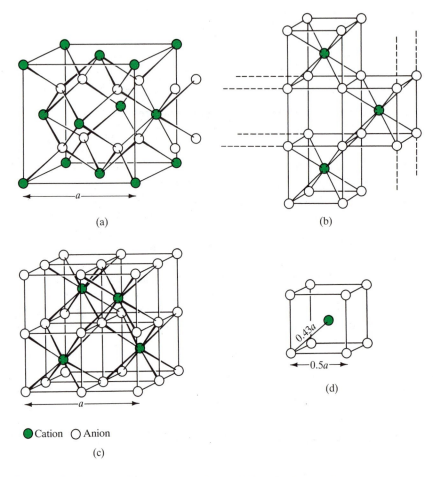

Figure 3.3   The crystal structure of fluorite MX_2. (a) Unit cell as a ccp array of cations. (b) and (c) The same structure redrawn as a simple cubic array of anions: the unit cell is marked by a coloured outline in (c). (d) Cell dimensions.

The number of Schottky defects in a crystal of composition MX is given by:

$$n_S \approx N\exp(-\Delta H_S/2kT), \tag{3.1}$$

where n_S is the number of Schottky defects per m^3, at T K, in a crystal with N cation and N anion sites per m^3; ΔH_S is the enthalpy required to form one defect. It is quite a simple matter to derive these equations for the equilibrium concentration of Schottky defects by considering the change in entropy of a perfect crystal due to the introduction of defects. The change in entropy will be due to the vibrations of atoms around the defects and also to the arrangement of the defects. It is possible to estimate this latter quantity, the **configurational entropy**, using the methods of statistical mechanics.

If the number of Schottky defects is n_S per m^3 at TK, then there will be n_S cation vacancies and n_S anion vacancies in a crystal containing N possible cation sites and N possible anion sites per m^3. The **Boltzmann formula** tells us that the entropy of such a system is given by:

$$S = k \ln W, \tag{3.2}$$

where W is the number of ways of distributing n_S defects over N possible sites at random, and k is the Boltzmann constant ($1.380\,622 \times 10^{-23}\,J\,K^{-1}$). Probability theory shows that W is given by:

$$W = \frac{N!}{(N - n)!n!}, \tag{3.3}$$

where $N!$ is called 'factorial N' and is mathematical shorthand for:

$$N \times (N - 1) \times (N - 2) \times (N - 3)\ldots \times 1.$$

So, the number of ways we can distribute the cation vacancies will be

$$W_c = \frac{N!}{(N - n_S)!n_S!}$$

and similarly for the anion vacancies

$$W_a = \frac{N!}{(N - n_S)!n_S!}.$$

The total number of ways of distributing these defects, W, is given by the product of W_c and W_a:

$$W = W_c W_a$$

and the change in entropy due to introducing the defects into a perfect crystal is thus:

$$\Delta S = k \ln W = k \ln\left(\frac{N!}{(N - n_S)!n_S!}\right)^2$$

$$= 2k \ln \frac{N!}{(N - n_S)!n_S!}.$$

We can simplify this expression using **Stirling's approximation** that

$$\ln N! \approx N \ln N - N$$

and the expression becomes (after manipulation)

$$\Delta S = 2k \{N \ln N - (N - n_S)\ln(N - n_S) - n_S \ln n_S\}.$$

If the enthalpy change for the formation of a single defect is ΔH_S and we assume that the enthalpy change for the formation of n_S defects is $n_S\Delta H_S$, then the Gibbs free energy change is given by:

$$\Delta G = n_S\Delta H_S - 2kT\{N\ln N - (N - n_S)\ln(N - n_S) - n_S\ln n_S\}.$$

At equilibrium, at constant T, the Gibbs free energy of the system must be a minimum with respect to changes in the number of defects n_S; thus

$$\left(\frac{d\Delta G}{dn_S}\right) = 0.$$

So,

$$\Delta H_S - 2kT\frac{d}{dn_S}[N\ln N - (N - n_S)\ln(N - n_S) - n_S\ln n_S] = 0.$$

$N\ln N$ is a constant and hence its differential is zero; the differential of $\ln x$ is $1/x$ and of $(x\ln x)$ is $(1 + \ln x)$. On differentiating we get:

$$\Delta H_S - 2kT[\ln(N - n_S) + 1 - \ln n_S - 1] = 0,$$

hence,

$$\Delta H_S = 2kT\ln\left[\frac{(N - n_S)}{n_S}\right]$$

and

$$n_S = (N - n_S)\exp(-\Delta H_S/2kT)$$

as $N \gg n_S$ we can approximate $(N - n_S)$ by N, finally giving:

$$\boxed{n_S = N\exp(-\Delta H_S/2kT)} \tag{3.1}$$

If we express this equation in molar quantities it becomes:

$$n_S \approx N\exp(-\Delta H_S/2RT), \tag{3.4}$$

where now ΔH_S is the enthalpy required to form one mole of Schottky defects and R is the gas constant, $8.314\,J\,K\,mol^{-1}$: the units of ΔH_S are $J\,mol^{-1}$.

By a similar analysis we find that the number of Frenkel defects present in a crystal MX is given by the expression:

$$\boxed{n_F = (NN_i)^{\frac{1}{2}}\exp(-\Delta H_F/2kT)} \tag{3.5}$$

where n_F is the number of Frenkel defects per m^3, N is the number of lattice sites and N_i the number of interstitial sites available. ΔH_F is the

Table 3.1 The formation enthalpy of Schottky and Frenkel defects in some selected compounds

	Compound	$\Delta H(10^{-19} J)$	$\Delta H(eV)^*$
Schottky defects	MgO	10.57	6.60
	CaO	9.77	6.10
	LiF	3.75	2.34
	LiCl	3.40	2.12
	LiBr	2.88	1.80
	LiI	2.08	1.30
	NaCl	3.69	2.30
	KCl	3.62	2.26
Frenkel defects	UO_2	5.45	3.40
	ZrO_2	6.57	4.10
	CaF_2	4.49	2.80
	SrF_2	1.12	0.70
	AgCl	2.56	1.60
	AgBr	1.92	1.20
	β-AgI	1.12	0.70

*The literature often quotes values in eV, so these are included for comparison: $1\,eV = 1.60219 \times 10^{-19}\,J$.

enthalpy of formation of one Frenkel defect. If ΔH_F is the enthalpy of formation of one mole of Frenkel defects the expression becomes:

$$n_F = (N N_i)^{\frac{1}{2}} \exp(-\Delta H_F/2RT). \tag{3.6}$$

Table 3.1 lists some enthalpy of formation values for Schottky and Frenkel defects in various crystals.

Using the information in Table 3.1 and equation (3.1) we can now get an idea of how many defects are present in a crystal. Assume that ΔH_S has a middle of the range value of $5 \times 10^{-19}\,J$. Substituting in equation (3.1) we find that the proportion of vacant sites n_s/N at 300 K is 6.12×10^{-27}; at 1000 K this rises to 1.37×10^{-8}. This shows what a low concentration of Schottky defects is present at room temperature. Even when the temperature is raised to 1000 K we still find only of the order of one or two vacancies per hundred million sites!

Whether Schottky or Frenkel defects are found in a crystal depends in the main on the value of ΔH, the defect with the lower ΔH value predominating. In some crystals it is possible for both types of defect to be present.

We will see in a later section that in order to change the properties of crystals, particularly their ionic conductivity, we may wish to introduce more defects into the crystal. It is important, therefore, at this stage to consider how this might be done.

Table 3.2   Values of n_S/N

$T(K)$	$\Delta H_S = 5 \times 10^{-19}\,J$	$\Delta H_S = 1 \times 10^{-19}\,J$
300	6.12×10^{-27}	5.72×10^{-6}
1000	1.37×10^{-8}	2.67×10^{-2}

First, we have seen from the above calculation, that raising the temperature introduces more defects. We would have expected this to happen because defect formation is an endothermic process and **Le Chatelier's principle** tells us that increasing the temperature of an endothermic reaction will favour the products, in this case defects. Second, if it were possible to decrease the enthalpy of formation of a defect, ΔH_S or ΔH_F, this would also increase the proportion of defects present. A simple calculation, again using equation (3.1), but now with a lower value for ΔH_S, say $1 \times 10^{-19}\,J$ allows us to see this. Table 3.2 compares the results. This has had a dramatic effect on the numbers of defects! At 1000 K we are now finding of the order of three in every hundred. It is difficult to see how the value of ΔH could be manipulated within a crystal, but we do find crystals where the value of ΔH is lower than usual due to the nature of the structure, and this can be exploited. This is true for one of the systems that we shall look at in detail later, α-AgI. Third, if we introduce impurities into a crystal selectively, we can increase the defect population.

3.2.3 Extrinsic defects

We can introduce vacancies into a crystal by **doping** it with a selected impurity: for instance if we add $CaCl_2$ to a NaCl crystal, each Ca^{2+} ion replaces two Na^+ ions in order to preserve electrical neutrality, and so one cation vacancy is created. Such created vacancies are known as **extrinsic**. An important example found later in this chapter is that of **zirconia (ZrO_2)**; this structure can be stabilized by doping with CaO, when the Ca^{2+} ions replace the Zr(IV) atoms in the lattice. The charge compensation in this case is achieved by the production of anion vacancies on the oxide sublattice.

3.3 IONIC CONDUCTIVITY IN SOLIDS

One of the most important aspects of point defects is that they make it possible for atoms or ions to move through the structure. If a crystal structure were perfect it would be difficult to envisage how the movement

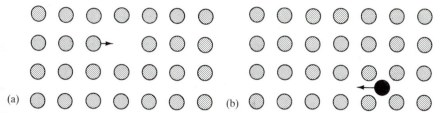

Figure 3.4 Schematic representation of ionic motion by (a) a vacancy mechanism and (b) an interstitial mechanism.

of atoms, either **diffusion** through the lattice or **ionic conductivity** (ion transport under the influence of an external electric field), could take place. Setting up equations to describe either diffusion or conductivity in solids is a very similar process, and so we have chosen to concentrate here on conductivity, because many of the examples later in the chapter are of solid electrolytes.

Two possible mechanisms for the movement of ions through a lattice are sketched in Figure 3.4. In (a) an ion hops or jumps from its normal position on the lattice to a neighbouring equivalent but vacant site. This is called the **vacancy mechanism** (it can equally well be described as the movement of a vacancy rather than the movement of the ion). Figure 3.4b shows both the vacancy and an **interstitial mechanism** where an interstitial ion jumps or hops to an adjacent equivalent site. These simple pictures of movement in an ionic lattice are known as the **hopping model**, and ignore more complicated cooperative motions.

Ionic conductivity, σ, is defined in the same way as electronic conductivity:

$$\sigma = nZe\mu, \tag{3.7}$$

where n is the number of charge carriers per unit volume, Ze is their charge (expressed as a multiple of the charge on an electron, $e = 1.602\,189 \times 10^{-19}$ C), and μ is their **mobility**, which is a measure of the drift velocity in a constant electric field. Table 3.3 shows the sort of conductivity values one might expect to find for different materials. As we might expect, ionic crystals, although they can conduct, are poor conductors compared with metals. This is a direct reflection of the difficulty the charge-carrier (in this case an ion, although sometimes an electron) has in moving through the crystal lattice.

Equation (3.7) is a general equation defining conductivity in all conducting materials. In order to understand why some ionic solids conduct better than others it is useful to look at the definition more closely in terms of the hopping model that we have set up. First of all we have said that an electric current is carried in an ionic solid by the defects. In the

Table 3.3 Typical values of electrical conductivity

	Material	Conductivity (Sm^{-1})
Ionic conductors	Ionic crystals	$<10^{-16}-10^{-2}$
	Solid electrolytes	$10^{-1}-10^{3}$
	Strong (liquid) electrolytes	$10^{-1}-10^{3}$
Electronic conductors	Metals	$10^{3}-10^{7}$
	Semiconductors	$10^{-3}-10^{4}$
	Insulators	$<10^{-10}$

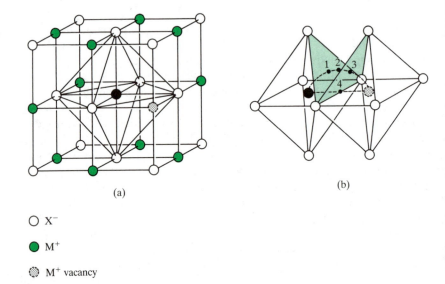

○ X^{-}

● M^{+}

◉ M^{+} vacancy

Figure 3.5 Sodium chloride type structure, showing (a) coordination octahedron of central cation, (b) coordination octahedra of central cation and adjacent vacancy.

cases of crystals where the ionic conductivity is carried by the vacancy or interstitial mechanism, n, the concentration of charge carriers, will be closely related to the concentration of defects in the crystal, n_S or n_F. μ will thus refer to the mobility of these defects in such cases.

Let us look more closely at the mobility of the defects. Take the case of NaCl which contains Schottky defects. The Na^{+} ions are the ones which move because these are the smallest; however, even these meet quite a lot of resistance (Figure 3.5). We have used dotted lines to illustrate two possible routes that the Na^{+} ion could take from the centre of the unit cell to an adjacent vacant site. The direct route (labelled 4) is clearly going to be very unlikely as it leads directly between two Cl^{-} ions

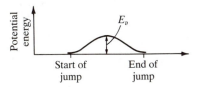

Figure 3.6 Schematic representation of the change in energy during motion of an ion along the lowest energy path.

which in a close-packed structure such as this are going to be very close together. The other pathway first passes through one of the triangular faces of the octahedron (point 1), then through one of the tetrahedral holes (point 2), and finally through another triangular face (point 3), before finally arriving at the vacant octahedral site. The coordination of the Na^+ ion will change from $6 \rightarrow 3 \rightarrow 4 \rightarrow 3 \rightarrow 6$ as it jumps from one site to the other. While there is clearly going to be an energy barrier to this happening, it will not be as large as for the direct path where the Na^+ ion becomes two-coordinate. In general we would expect the ion to follow the lowest energy path available; a schematic diagram of the energy changes involved in such a pathway is shown in Figure 3.6. Notice that the energy of the ion is the same at the beginning and end of the jump; the energy required to make the jump, E_a, is known as the **activation energy** for the jump. This means that the temperature dependence of the mobility of the ions can be expressed by an **Arrhenius equation**:

$$\mu \propto \exp(-E_a/kT) \tag{3.8}$$

or,

$$\mu = \mu_0 \exp(-E_a/kT), \tag{3.9}$$

where μ_0 is a proportionality constant known as a pre-exponential factor. μ_0 depends on several factors: the number of times per second that the ion attempts the move, v, called the **attempt frequency** (this is a frequency of vibration of the lattice, of the order of $10^{12}-10^{13}$ Hz); the distance moved by the ion; and the size of the external field. If the external field is small (up to about $300\,V\,cm^{-1}$) a temperature dependence of $1/T$ is introduced into the pre-exponential factor.

If we combine all this information in equation (3.7), we arrive at an expression for the variation of ionic conductivity with temperature that has the form:

$$\sigma = (\sigma_0/T) \exp(-E_a/T) . \tag{3.10}$$

The term σ_0 now contains n and Ze, as well as the information on attempt frequency and jump distance. This expression accounts for the fact that

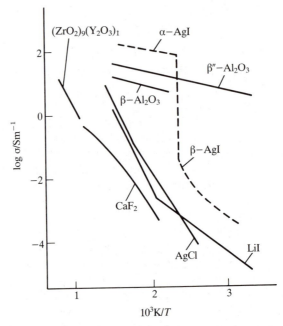

Figure 3.7 The conductivities of selected solid electrolytes over a range of temperatures.

ionic conductivity increases with temperature. If we now take logs of equation (3.10) we get:

$$\ln \sigma T = \ln \sigma_0 - E_a/T.$$

Plotting $\ln \sigma T$ against $1/T$ should produce a straight line with a slope of $-E_a$. The expression in equation (3.10) is sometimes plotted empirically as:

$$\sigma = \sigma_0 \exp(-E_a/T)$$

because plotting either $\ln \sigma T$ or $\ln \sigma$ makes little difference to the slope; both types of plot are found in the literature. The results of doing this for several compounds are shown in Figure 3.7.

Ignore the AgI line for the moment which we discuss in the next section. The other lines on the plot are straight lines apart from the one for LiI where there are clearly two lines of differing slope. So it appears as though the model we have set up does describe the behaviour of many systems. But how about LiI? In fact other crystals also show this kink in the plot (some experimental data is shown for NaCl in Figure 3.8, where

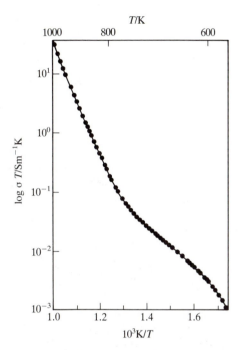

Figure 3.8 The ionic conductivity of NaCl plotted against reciprocal temperature.

it can also be seen). Is it possible to explain this using the equations that we have just set up? The answer is yes.

The explanation for the two slopes in the plot lies in the fact that even a very pure crystal of NaCl contains some impurities and the line corresponding to low temperatures (on the right of the plot) is due to the **extrinsic vacancies**: at low temperatures the concentration of **intrinsic vacancies** is so small that it can be ignored because it is dominated by the defects created by the impurity. For a particular amount of impurity the number of vacancies present will be essentially constant. μ in this **extrinsic region** thus depends only on the cation mobility due to these extrinsic defects, whose temperature dependence is given by equation (3.9):

$$\mu = \mu_0 \exp(-E_a/kT). \tag{3.9}$$

However, at the higher temperatures on the left-hand side of the graph, the concentration of intrinsic defects has increased to such an extent that it now is similar to or greater than the concentration of extrinsic defects. The concentration of intrinsic defects, unlike that of the extrinsic defects will not be constant; indeed it will vary according to equation (3.1), as we showed earlier:

$$n_S = N \exp(-\Delta H_S/2kT) \tag{3.1}$$

and so the conductivity in this **intrinsic region** on the left-hand side of the plot is given by:

$$\sigma = \frac{\sigma'}{T} \exp(-E_a/kT) \exp(-\Delta H_S/2kT). \tag{3.11}$$

A plot of $\ln \sigma T$ versus $1/T$ in this case will give a greater value for the activation energy, E_S, because it will actually depend on two terms: the activation energy for the cation jump, E_a, and the enthalpy of formation of a Schottky defect:

$$E_S = E_a + \tfrac{1}{2}\Delta H_S. \tag{3.12}$$

Similarly for a system with Frenkel defects:

$$E_F = E_a + \tfrac{1}{2}\Delta H_F. \tag{3.13}$$

From plots such as these we find that the activation energies lie in the range 0.05--1.1 eV, rather lower than the enthalpies of defect formation. As we have seen, raising the temperature increases the number of defects, and so increases the conductivity of a solid. Better than increasing the temperature to increase conductivity is to find materials that have low activation energies, less than about 0.2 eV. We find such materials in the top right-hand corner of Figure 3.7.

3.4 SOLID ELECTROLYTES

Much of the recent interest and research in solid state chemistry is related to the electrical conductivity properties of solids; new electrochemical cells and batteries are being developed that contain solid, rather than liquid, electrolytes.

A battery is an electrochemical cell that produces an electric current at a constant voltage as a result of a chemical reaction. The **emf** or voltage produced by the cell under standard open circuit conditions is related to the standard Gibb's free energy change for the reaction by the following equation:

$$\Delta G^{\ominus} = -nE^{\ominus}F, \tag{3.14}$$

where n is the number of electrons transferred in the reaction, $E^{\ominus}$ is the standard electromotive force (emf) of the cell (the voltage delivered under standard conditions), and F is the Faraday ($96\,485\,C\,mol^{-1}$ or $96\,485\,J\,V^{-1}$).

The ions taking part in the reaction pass through an **electrolyte** and are then either oxidized or reduced at an **electrode**; the electrons are released at the positively charged electrode, the **anode**, and pass around an external circuit to the **cathode**. The electric current in the external circuit can be harnessed to do useful work.

The reason that these solid state batteries are potentially useful is that they can perform over a wide temperature range, they have a long shelf life and it is possible to manufacture them so that they are extremely small. There is also interest in solid electrolytes in the production of **secondary** or **storage batteries**; these batteries are reversible because once the chemical reaction has taken place the reactant concentrations can be restored by reversing the cell reaction using an external source of electricity. Once a **primary battery** has discharged the reaction cannot be reversed and it is useless. If storage batteries can be made to give sufficient power and they are light, they become useful as alternative fuel sources to power cars for instance; we will look in some detail at the development of the sodium–sulphur battery for this purpose.

To take a simple example, LiI, although it has a fairly low ionic conductivity (Figure 3.7), is used in the manufacture of the batteries that are used in heart pace-makers: LiI is the solid electrolyte that separates the anode (lithium) from the cathode (iodine embedded in a conducting polymer). The cell reaction is:

anode: $2Li(s) = 2Li^+ + 2e^-$
cathode: $I_2(s) + 2e^- = 2I^-$
overall: $2Li(s) + I_2(s) = 2LiI(s)$

Because the LiI contains Schottky defects, the small Li^+ cations are able to pass through the solid electrolyte, while the released electrons go round an external circuit.

3.4.1. Fast-ion conductors

Some ionic solids have been discovered that have a much higher conductivity than is typical for such compounds and these are known as **fast-ion conductors**. One of the earliest to be noticed, in 1913 by Tubandt and Lorenz, was a high temperature phase of silver iodide.

α-AgI

Below 146°C there are two phases of AgI: γ-AgI which has the zinc blende structure and β-AgI with the wurtzite structure; both are based on a close-packed array of iodide ions with half of the tetrahedral holes filled. However, above 146°C a new phase, α-AgI, is observed where the

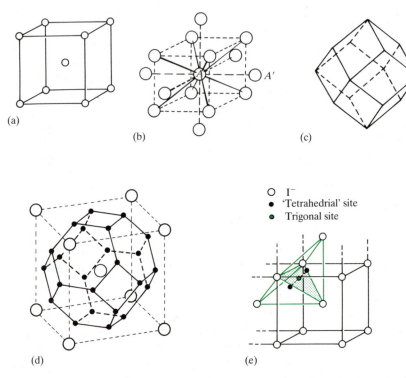

(a)

(b)

(c)

(d)

(e)

○ I⁻

● 'Tetrahedrial' site

● Trigonal site

Figure 3.9 Building up the structure of α-AgI. (a) The body-centred cubic array of I⁻ ions. (b) bcc array extended to next-nearest neighbours. (c) Rhombic dodecahedron. (d) bcc array with enclosed truncated octahedron. (e) The positions of two tetrahedral sites and a trigonal site between them (coloured).

iodine ions now have a body-centred cubic lattice. A dramatic increase in conductivity is observed for this phase (Figure 3.7): the conductivity of α-AgI is very high, $131\,\mathrm{S\,m^{-1}}$, a factor of 10^4 higher than that of β- or γ-AgI, comparable with the conductivity of the best conducting liquid electrolytes. How can we explain this startling phenomenon?

The explanation lies in the crystal structure of α-AgI. The structure is based on a body-centred cubic array of I⁻ ions (Figure 3.9a). Each I⁻ ion in the array is surrounded by eight equidistant I⁻ ions. In order to see where the Ag⁺ ions fit into the structure we need to look at the bcc structure in more detail. Figure 3.9b shows the same unit cell but with the next-nearest neighbours added in, these are the six at the body centres of the surrounding unit cells and are only 15% further away than the eight

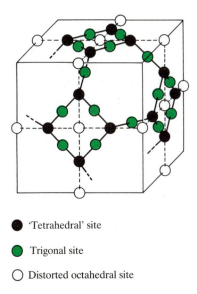

● 'Tetrahedral' site

● Trigonal site

○ Distorted octahedral site

Figure 3.10 Possible cation sites in the bcc structure of α-AgI. The thick solid and dashed lines mark possible diffusion paths.

immediate neighbours. This means that the atom marked A is surrounded by 14 other identical atoms although not in a completely regular way: these 14 atoms lie at the vertices of a **rhombic dodecahedron** (Figure 3.9c). It can be convenient to describe the structure in terms of the space-filling **truncated octahedron** shown in Figure 3.9d, which has six square faces and eight hexagonal faces corresponding to the two sets of neighbours. This is called the **domain** of an atom. Each vertex of the domain lies at the centre of an interstice (like the tetrahedral and octahedral holes found in the close-packed structures), which in this structure is a distorted tetrahedron. Figure 3.9e shows two such adjacent distorted tetrahedral holes, and the 'tetrahedral' site lying on the face of the unit cell. Where the two 'tetrahedra' join, they share a common triangular face; a trigonal site in the centre of this face is also marked in Figure 3.9e. A third type of site in the centre of each face and of each edge of the unit cell can also be defined; these sites have distorted octahedral coordination. The structure thus possesses the unusual feature that there are a great variety of positions that can be adopted by the Ag^+ ions, and Figure 3.10 sketches many of them. Each AgI unit cell possesses two I^- ions ($(8 \times \frac{1}{8})$ at the corners and one at the body centre) and so positions for only two Ag^+ ions need be found. From the possible sites that we have described there are 6 distorted octahedral, 12 'tetrahedral' and 24 trigonal, a choice of 42 possible sites – the Ag^+ ions have a huge choice of position open to them! Structure determinations indicate that the Ag^+ ions are statistically distributed amongst the 12 tetrahedral sites all of which have the same energy; so, counting these sites alone, we find that there are *five* spare sites per Ag^+ ion. We can visualize the Ag^+ ions moving from tetrahedral site to tetrahedral site by jumping through the vacant trigonal site, following the paths marked with solid lines in Figure 3.10, continually creating and destroying Frenkel defects and able to move easily through the lattice. The paths marked with thin dotted lines in Figure 3.10 require higher energy as they pass through the vacant 'octahedral' sites which are more crowded. The jump that we have described only changes the co-ordination number from $4 \rightarrow 3 \rightarrow 4$ and experimentally the activation energy is found to be very low, $0.05\,eV$. This easy movement of the Ag^+ ions through the lattice has often been described as a **molten sublattice** of Ag^+ ions.

The very high conductivity of α-AgI seems to arise because of a combination of favourable factors. The features that have contributed to this are listed below.

1. The charge on the ions is low; in AgI the mobile Ag^+ ions are monovalent.
2. The coordination around the ions is also low, so that when they jump from site to site, the coordination changes only by a little, affording a route through the lattice with a low activation energy.
3. The anion is rather **polarizable**; this means that the electron cloud surrounding it is easily distorted. This makes the passage of a cation past an anion easier.
4. There are a large number of vacant sites for the cations to move into.

These are properties that are known to be important when looking for other fast-ion conductors.

$RbAg_4I_5$

The special electrical properties of α-AgI inevitably led to a search for other solids exhibiting high ionic conductivity, preferably at temperatures lower than 146°C. The most successful so far (though there are many others) has been the partial replacement of Ag by Rb, forming the compound $RbAg_4I_5$. This compound has an ionic conductivity at room temperature of $25\,S\,m^{-1}$, with an activation energy of only $0.07\,eV$. The crystal structure is different from that of α-AgI, but similarly the Rb^+ and I^- ions form a rigid array while the Ag^+ ions are randomly distributed over a network of tetrahedral sites through which they can move.

If a conducting ionic solid is to be useful as a solid electrolyte in a battery, not only must it possess a high conductivity, but it must also have *negligible electronic conductivity*. This is to stop the battery short-circuiting; the electrons must only pass through the external circuit, whence they can be harnessed for work. $RbAg_4I_5$ has been used as the solid electrolyte in batteries with electrodes made of Ag and RbI_3. Such cells operate over a wide temperature range (-55 to $+200°C$), have a long shelf life and can withstand mechanical shock.

Table 3.4 lists ionic conductors that behave in a similar way to α-AgI. Some of these structures are based on a close-packed array of anions and this is noted in the table; the conducting mechanism in these compounds is similar to that in α-AgI. The chalcogenide structures such as silver sulphide and selenide tend to show electronic conductivity as well as ionic, although this can be quite useful in an electrode material as opposed to an electrolyte.

Table 3.4 α-AgI related ionic conductors

Anion structure	bcc	fcc	hcp	Other
	α-AgI	α-CuI	β-CuBr	$RbAg_4I_5$
	α-CuBr	α-Ag$_2$Te		
	α-Ag$_2$S	α-Cu$_2$Se		
	α-Ag$_2$Se	α-Ag$_2$HgI$_4$		

Stabilized zirconia

The fluorite structure, as we know from Figure 3.3, has plenty of empty space which can enable a F^- ion to move into an interstitial site. If the activation energy for this process is low enough we might expect compounds with this structure to show ionic conductivity. Indeed PbF_2 has a low ionic conductivity at room temperature, but this increases smoothly with temperature to a limiting value of $\sim 500\,S\,m^{-1}$ at $500°C$.

Zirconia (ZrO_2), is found as the mineral baddeleyite which has a monoclinic structure. Above $1000°C$ this changes to a tetragonal form and at high temperatures ZrO_2 adopts the cubic fluorite structure. The cubic form can be stabilized at room temperature, however, by the addition of CaO, which forms a new phase with the ZrO_2 after heating at about $1600°C$. The phase diagram is shown in Figure 3.11 and shows that at about 15 mole % CaO the new phase, known as **calcia-stabilized zirconia**, appears alone and persists until about 28 mole % CaO. Above 28 mole % another new phase appears, finally appearing pure as $CaZrO_3$. If the Ca^{2+} ions sit on Zr^{4+} sites, compensating vacancies are created in the O^{2-} sublattice: thus for every Ca^{2+} ion taken into the structure *one* anion vacancy is created. The large proportions of CaO mean that such oxides have a huge defect population of oxygen in the structure. Consequently these materials are exceptionally good fast-ion conductors of O^{2-} anions.

There are many materials made with this type of structure based on other oxides such as ThO_2 and HfO_2 as well as ZrO_2, and doped with oxides of the lanthanide series as well as CaO; these are also collectively

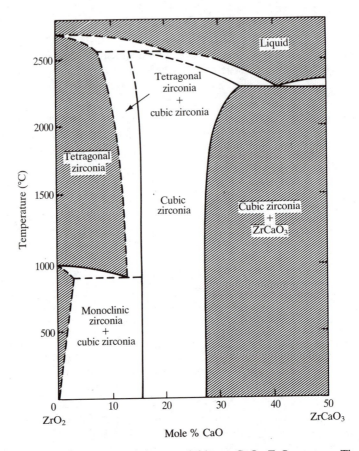

Figure 3.11 Phase diagram of the pseudobinary $CaO-ZrO_2$ system. The cubic calcia-stabilized zirconia phase occupies the central band in the diagram and is stable to about 2400°C.

known as stabilized zirconias. They are widely used in electrochemical systems.

One of the most interesting applications is the use of calcia-stabilized zirconia in both **oxygen meters** and **oxygen sensors**, which are based on a specialized electrochemical cell; they work in the following way. Figure 3.12 shows a slab of calcia-stabilized zirconia acting as a solid electrolyte that separates two regions containing oxygen at different pressures. Gas pressures tend to equalize if they can and so, if $p' > p''$, oxygen ions, which are able to pass through the stabilized zirconia, tend to pass through the solid from the left-hand side to the right. This tendency produces a potential difference (because the ions are charged), indicating

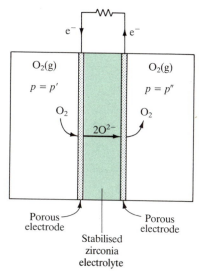

Figure 3.12 Schematic representation of an oxygen meter.

that oxygen is present (in the sensor) and measurement of this potential gives a measure of the oxygen pressure difference (in the oxygen meter).

Oxygen gas is reduced to O^{2-} at the left-hand electrode (LHE), and the oxide ions are able to pass through the doped zirconia and are re-oxidized to oxygen gas at the right-hand electrode (RHE). The equation for the cell reaction is as follows:

LHE: $\quad O_2(p') + 4e \rightarrow 2O^{2-},$ (3.15)

RHE: $\quad 2O^{2-} \rightarrow O_2(p'') + 4e,$ (3.16)

overall: $\quad O_2(p') \rightarrow O_2(p'').$ (3.17)

Under standard conditions we can relate the change in Gibb's free energy for the above reaction to the standard emf of the cell:

$$\Delta G^{\ominus} = -nE^{\ominus}F.$$ (3.14)

The **Nernst equation** allows calculation of the cell emf under non-standard conditions, E. If the cell reaction is given by a general equation:

$$aA + bB + \ldots + ne = xX + yY + \ldots$$ (3.18)

then,

$$E = E^{\ominus} - \left(\frac{2.303RT}{nF}\right) \log \left\{\frac{a_X^x a_Y^y \ldots}{a_A^a a_B^b \ldots}\right\}$$ (3.19)

where the quantities a_X, etc. are the **activities** of the reactants and products. Applying the Nernst equation to the cell reaction in the oxygen meter we get:

$$E = E^{\oplus} - \left(\frac{2.303RT}{4F}\right)\log\left(\frac{p''}{p'}\right). \tag{3.20}$$

In this particular case $E^{\oplus}$ is actually zero because under standard conditions the pressure of the oxygen on each side would be 1 atm and there would be no net potential difference. The pressure of the oxygen on one side of the cell (say p'') is set to be a known reference pressure, usually either pure oxygen at 1 atm or atmospheric oxygen pressure ($\sim$0.21 atm). Making these two changes we get:

$$E = \left(\frac{2.303RT}{4F}\right)\log\left(\frac{p'}{p_{ref}}\right). \tag{3.21}$$

Clearly, all the quantities in this equation are now known or can be measured, affording a direct measure of the unknown oxygen pressure, p'. For this cell to operate there must, as we stipulated before, be no electronic conduction through the electrolyte.

Oxygen meters find industrial uses in the detection of oxygen in waste gases from chimneys or exhaust pipes, in investigation into the operation of furnaces, and even in measuring the oxygen content in molten metals during their production.

β-Alumina

β-Alumina is the name given to a series of compounds which show fast-ion conducting properties. The parent compound is sodium β-alumina, $Na_2O.11Al_2O_3$ ($NaAl_{11}O_{17}$) and is found as a by-product from the glass industry. (The compound was originally thought to be a polymorph of Al_2O_3 and was named as such; it was only later found to contain sodium ions! However, the original name has stuck.) The general formula for the series is $M_2O.nX_2O_3$, where n can range from 5 to 11; M is a monovalent cation such as (alkali metal)$^+$, Cu^+, Ag^+, NH_4^+, and X is a trivalent cation: Al^{3+}, Ga^{3+} or Fe^{3+}.

The real composition of β-alumina actually varies quite considerably from the ideal formula and the materials are always found to be rich in Na^+ and O^{2-} ions to a greater or lesser extent. Two modifications of the structure exist, β_ and β''_, depending on the number of Na^+ ions present; β''_ occurs in the more soda-rich crystals where $n = 5$–7 and β_ occurs for $n = 8$–11.

Interest in these compounds started in 1966 when research at the Ford Motor Co. showed that the Na^+ ions were very mobile both at room temperature and above. The high conductivity of the Na^+ ions in

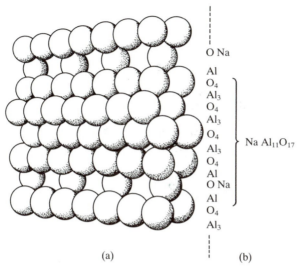

O Na

Al
O_4
Al_3
O_4
Al_3
O_4
Al_3
O_4
Al
O Na
Al
O_4
Al_3

$\left.\right\}$ Na $Al_{11}O_{17}$

(a)   (b)

Figure 3.13 (a) Oxide layers in β-alumina. (b) The ratio of atoms in each layer of the structure.

β-alumina is due to the crystal structure. This consists of close-packed layers of oxide ions but every fifth layer has three-quarters of the oxygens missing (Figure 3.13). The four close-packed layers contain the Al^{3+} ions in both octahedral and tetrahedral holes [they are known as the 'spinel blocks' due to their similarity to the crystal structure of the mineral spinel $MgAl_2O_4$ (discussed in Chapter 1 and illustrated in Figure 1.34)]. The groups of four close-packed oxide layers are held apart by a rigid Al–O–Al linkage; this O atom constituting the fifth oxide layer which contains only a quarter of the number of oxygens of each of the other layers. The Na^+ ions are found in these fifth oxide layers, which are mirror planes in the structure. The overall stoichiometry of the structure is shown layer by layer in Figure 3.13b: the sequence of layers is

C (ABCA) B (ACBA) C . . . ,

where the parentheses enclose the four close-packed layers and the intermediate symbols refer to the fifth oxide layer. The structure of β″-alumina is not shown here; it is similar, but differs in the stacking sequences of the close-packed layers:

C (ABCA) B (CABC) A (BCAB) C . . . .

The crystal structure of β-alumina is shown in Figure 3.14. The Na^+ ions are located in the faces at the top and bottom of the half-cell shown. They can move around easily for two reasons: (i) there are plenty of vacancies, so there is a choice of sites, and (ii) the Na^+ ions are smaller than the

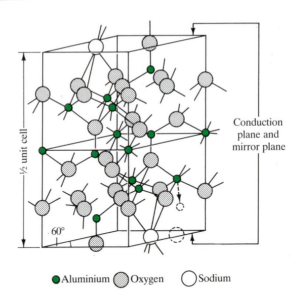

Conduction
plane and
mirror plane

½ unit cell

60°

● Aluminium   ◯ Oxygen   ◯ Sodium

Figure 3.14 Structure of half the unit cell of stoichiometric sodium β-alumina. (The dashed lines indicate the effect of incorporating additional oxygen in the conduction plane.)

O^{2-} ions. Conduction in the β-aluminas only occurs within the planes containing the oxygen vacancies; these are known as the **conduction planes**. The alkali metal cations cannot penetrate the dense 'spinel blocks', but can move easily from site to site within the plane. As stated previously, β-alumina is never found in the stoichiometric form; it is always Na_2O rich. The sodium-rich compounds have a much higher conductivity than stoichiometric β-alumina. The extra sodium ions have to be compensated by a counter defect in order to keep the overall charge on the compound at zero. There is more than one possibility for this but in practice it is found that extra oxide ions provide the compensation and the overall formula can be written as $(Na_2O)_{1+x}.11Al_2O_3$. The extra sodium and oxide ions both occupy the fifth oxide layer; the O^{2-} ions are locked into position by an Al^{3+} moving out from the spinel block (Figure 3.14), and the Na^+ ions become part of the mobile pool of ions. The Na^+ ions are so fluid that the ionic conductivity in β-alumina at 300°C is close to that of typical liquid electrolytes at ambient temperature.

In general, the β″-aluminas have even higher conductivities than the β-aluminas but tend to be more sensitive to moisture. In practice both types of solids are used as electrolytes in the manufacture of electrochemical cells. β-Alumina is used as an electrolyte in cells which are primarily intended as power supplies. From equation (3.14)

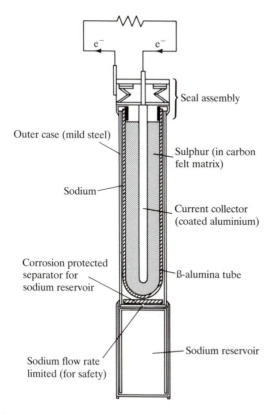

Figure 3.15 Schematic representation of a 'central sulphur' cell. Typically, the electrolyte tube is some 300 mm long and 30 mm in diameter.

$$\Delta G^{\ominus} = -nE^{\ominus}F \qquad (3.14)$$

we know that in order to obtain a large voltage from a cell we must have a cell reaction with a large negative Gibbs free energy change, such as a reaction between an alkali metal and a halogen might give, for instance. In terms of energy output per unit weight such a reaction can yield about 600 kW per hour per kg of material. Such reactive materials have to be separated by an electrolyte impermeable to electrons but which can be crossed by ions, and β-alumina has proved very useful in these cells.

The **sodium–sulphur** cell uses Na^+ β-alumina as the electrolyte (Figure 3.15). It has a high energy to mass ratio (1030 W h kg^{-1}) and is being developed for uses such as electric cars. The electrolyte separates molten sodium from molten sulphur which is contained in a graphite felt and is in contact with a current collector connected to the external circuit. The cell operates at 300°C; the heat required to maintain the cell at this operating temperature being supplied by the cell reaction itself.

The cell operates in the following way. At the sodium–electrolyte interface, the sodium atoms each lose an electron and enter the Na^+ layers in the crystal:

$$2Na(l) = 2Na^+ + 2e. \tag{3.22}$$

At the sulphur–electrolyte interface a complex reaction forming poly-sulphides of sodium takes place. In the early stages of discharge one such reaction would be:

$$2Na^+ + 5S(l) + 2e = Na_2S_5(l). \tag{3.23}$$

The overall cell reaction is then:

$$2Na(l) + 5S(1) = Na_2S_5(l). \tag{3.24}$$

Later, low polysulphides are formed, and the discharge is terminated at a composition of about Na_2S_3. In spite of their complexity the electrode process can be reversed by applying a current from an external source.

There is much commercial interest in this cell because it incorporates a highly energetic reaction between light substances and so can provide a high **energy density**, and it is reversible so the cells can be recharged. The cells obviously have to incorporate stringent safety features because they contain highly reactive and corrosive chemicals at 300°C! In spite of the restrictions it is hoped that they will become commercially viable.

3.5 PHOTOGRAPHY

A photographic emulsion consists of very small crystallites of AgBr (or AgBr–AgI) dispersed in gelatin; this is usually supported on paper or thin plastic to form a film. The crystallites are usually small triangular or hexagonal platelets, known as **grains**. They are grown very carefully *in situ* with as few structural defects as possible and range in size from 0.05 to 2×10^{-6} m. During the photographic process light falling on the AgBr produces Ag atoms in some of the grains; these eventually form the dark parts of the negative. The grains which are affected by the light contain the so-called **latent image**. It is important for the grains to be free from structural defects such as dislocations and grain boundaries as these interfere with the deposition of the Ag atoms on the surface of the grains. However, the formation of the latent image is dependent on the presence of point defects.

AgBr has the NaCl or rock-salt crystal structure. However, unlike the alkali halides, which contain mainly Schottky defects, AgBr has been shown to contain mostly Frenkel defects, in the form of interstitial Ag^+ ions. For a grain to possess a latent image it need only have as few as four Ag atoms in a cluster on the surface. The formation of the clusters of Ag atoms is a complex process that is still not fully understood. However, it is thought to take place in several stages. The first stage is when light strikes one of the AgBr crystals and an electron is promoted from the valence band to the conduction band. The band gap of AgBr is 2.7 eV, so the light absorbed is visible from the extreme blue end of the spectrum. This electron eventually neutralizes one of the interstitial silver ions (Ag_i^+):

$$Ag_i^+ + e = Ag.$$

In the next stages this Ag atom speck has to grow into a cluster of atoms on the surface of the crystal. A possible sequence of events for this is:

$$Ag + e = Ag^-$$
$$Ag^- + Ag_i^+ = Ag_2$$
$$Ag_2 + Ag_i^+ = Ag_3^+$$
$$Ag_3^+ + e = Ag_3$$
$$Ag_3 + e = Ag_3^-$$
$$Ag_3^- + Ag_i^+ = Ag_4 + \ldots .$$

Notice that only the odd numbered clusters seem to interact with the electrons.

In reality the process is even more complex than this because emulsions made from pure AgBr are not sensitive enough and so they also contain **sensitizers** such as sulphur or organic dyes, which absorb light of longer wavelength than AgBr and so extend the spectral range. The sensitizers form traps for the photoelectrons on the surfaces of the grains; these electrons then transfer from an excited energy level of the sensitizer to the conduction band of AgBr.

The film containing the latent image is then treated with various chemicals to produce a lasting negative. First of all it is developed: a reducing agent such as an alkaline solution of hydroquinone is used to reduce the AgBr crystals to Ag. The clusters of Ag atoms act as a catalyst to this reduction process, and all the grains with a latent image are reduced to Ag. The process is rate controlled, so the grains that have not reacted with the light are unaffected by the developer unless the film is developed for a very long time, when eventually they will be reduced and a fogged picture results. The final stage in producing a negative is to dissolve out the remaining light-sensitive AgBr. This is done using 'hypo'-sodium thiosulphate ($Na_2S_2O_3$), which forms a water-soluble complex with Ag^+ ions.

3.6 COLOUR CENTRES

During early research in Germany it was noticed that if crystals of the alkali halides were exposed to X-rays, they became brightly coloured.

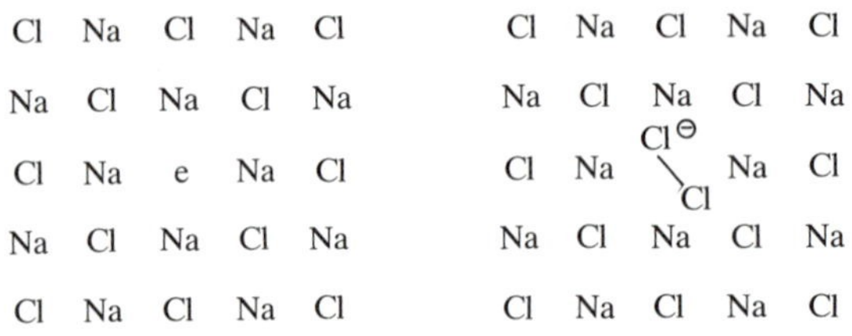

```
Cl  Na  Cl  Na  Cl          Cl  Na  Cl  Na  Cl

Na  Cl  Na  Cl  Na          Na  Cl  Na  Cl  Na
                                    Cl⊖
Cl  Na  e   Na  Cl          Cl  Na    \     Na  Cl
                                        Cl
Na  Cl  Na  Cl  Na          Na  Cl  Na  Cl  Na

Cl  Na  Cl  Na  Cl          Cl  Na  Cl  Na  Cl

            (a)                         (b)
```

Figure 3.16 (a) The F-centre, an electron trapped on an anion vacancy. (b) H-centre.

The colour was thought to be associated with a defect known then as a **Farbenzentre** (colour centre), now abbreviated to **F-centre**. Since then it has been found that many forms of high energy radiation (UV, X-rays, neutrons) will cause F-centres to form. The colour produced by the F-centres is always characteristic of the host crystal, so for instance, NaCl becomes deep yellowy-orange, KCl, violet and KBr, blue-green.

Subsequently it was found that F-centres can also be produced by heating a crystal in the vapour of an alkali metal; this gives a clue to the nature of these defects. The excess alkali metal atoms diffuse into the crystal and settle on cation sites; at the same time an equivalent number of anion site vacancies are created and ionization gives an alkali metal cation with an electron trapped at the anion vacancy (Figure 3.16). In fact, it does not even matter which alkali metal is used; if NaCl is heated with potassium the colour of the F-centre does not change because it is characteristic of the electron trapped at the anion vacancy in the host halide. Work with electron spin resonance (ESR) techniques has confirmed that F-centres are indeed unpaired electrons trapped at vacant lattice (anion) sites.

The trapped electron provides a classic example of an 'electron in a box'. A series of energy levels are available for the electron, and the energy required to transfer from one level to another falls in the visible part of the electromagnetic spectrum, hence the colour of the F-centre. There is an interesting natural example of this phenomenon: the mineral fluorite, CaF_2, is found in Derbyshire as 'Blue John' and its beautiful blue-purple coloration is due to the presence of F-centres.

Many other colour centres have now been characterized in alkali halide crystals. The **H-centre** is formed by heating (for instance) NaCl in Cl_2 gas. In this case a $[Cl_2]^-$ ion is formed and occupies a single anion site (Figure 3.16b). F-centres and H-centres are perfectly complementary, if they meet they cancel one another out!

Another interesting natural example of colour centres lies in the colour of smoky quartz and amethyst. These semiprecious stones are basically crystals of silica (SiO_2) with some impurity present. In the case of smoky quartz the silica contains a little Al impurity. The Al substitutes for the Si in the lattice and the electrical neutrality is maintained by H^+ present in the same amount as Al. The colour centre arises when ionizing radiation interacts with an $[AlO_4]^{5-}$ group liberating an electron which is then trapped by H^+:

$$[AlO_4]^{5-} + H^+ = [AlO_4]^{4-} + H. \tag{3.25}$$

The $[AlO_4]^{4-}$ group is now electron-deficient and can be considered to have a 'hole' trapped at its centre. This group is the colour centre, absorbing light and producing the smoky colour. In crystals of amethyst, the impurity present is Fe^{3+}, and on irradiation $[FeO_4]^{4-}$ colour centres are produced which absorb light to give the characteristic purple coloration.

3.7 NON-STOICHIOMETRIC COMPOUNDS

3.7.1 Introduction

Previous sections in this chapter have shown that it is possible to *introduce* defects into a perfect crystal by adding an impurity. Such an addition causes point defects of one sort or another to form but they no longer occur in complementary pairs. Such impurity-induced defects are said to be **extrinsic**. We have also noted that when assessing what defects have been created in a crystal, it is important to remember that the overall charge on the crystal must always be zero.

Colour centres are formed if a crystal of NaCl is heated in sodium vapour; sodium is taken into the crystal, and the formula becomes $Na_{1+x}Cl$. The sodium atoms occupy cation sites, creating an equivalent number of anion vacancies; they subsequently ionize to form a sodium cation with an electron trapped at the anion vacancy. The solid so formed is a **non-stoichiometric compound** because the ratio of the atomic components is no longer the simple integer that we have come to expect for well-characterized compounds. A careful analysis of many substances, particularly inorganic solids, shows that it is common for the atomic ratios to be non-integral: uranium dioxide for instance can range in composition from $UO_{1.65}$ to $UO_{2.25}$, certainly not the perfect UO_2 that we might expect! There are many other examples, some of which are discussed in some detail.

What kind of compounds are likely to be non-stoichiometric? 'Normal' covalent compounds are assumed to have a fixed composition where the atoms are usually held together by strong covalent bonds formed by the pairing of two electrons. Breaking these bonds usually takes quite a lot of energy, and so under normal circumstances a particular compound does not show a wide range of composition; this is true for most molecular organic compounds for instance. Ionic compounds also are usually stoichiometric, because to remove or add ions requires a considerable amount of energy. We have seen, however, that it is possible to make ionic crystals non-stoichiometric by doping them with an impurity, as with the example of Na added to NaCl. There is also another mechanism whereby ionic crystals can become non-stoichiometric: if the crystal contains an element with a variable valency, then a change in the number of ions of that element can be compensated by changes in ion charge; this maintains the charge balance but alters the stoichiometry. Elements with a variable valency mostly occur in the transition elements, the lanthanides and the actinides.

In summary, non-stoichiometric compounds can have formulae that do not have simple integer ratios of atoms; they also usually exhibit a range of composition. They can be made by introducing impurities into a system, but are frequently a consequence of the ability of the metal to exhibit variable valency. Table 3.5 lists a few non-stoichiometric compounds together with their composition ranges.

Until recently, defects both in stoichiometric and non-stoichiometric crystals were treated entirely from the point of view that point defects are

Table 3.5 Approximate composition ranges for some non-stoichiometric compounds

Compound		Composition range*
TiO_x	$[\approx TiO]$	$0.65 < x < 1.25$
	$[\approx TiO_2]$	$1.998 < x < 2.000$
VO_x	$[\approx VO]$	$0.79 < x < 1.29$
Mn_xO	$[\approx MnO]$	$0.848 < x < 1.000$
Fe_xO	$[FeO]$	$0.833 < x < 0.957$
Co_xO	$[\approx CoO]$	$0.988 < x < 1.000$
Ni_xO	$[\approx NiO]$	$0.999 < x < 1.000$
CeO'_x	$[\approx Ce_2O_3]$	$1.50 < x < 1.52$
ZrO_x	$[\approx ZrO_2]$	$1.700 < x < 2.004$
UO_x	$[\approx UO_2]$	$1.65 < x < 2.25$
$Li_xV_2O_5$		$0.2 < x < 0.33$
Li_xWO_3		$0 < x < 0.50$
TiS_x	$[\approx TiS]$	$0.971 < x < 1.064$
Nb_xS	$[\approx NbS]$	$0.92 < x < 1.00$
Y_xSe	$[\approx YSe]$	$1.00 < x < 1.33$
V_xTe_2	$[\approx VTe_2]$	$1.03 < x < 1.14$

* Note that all composition ranges are temperature-dependent and the figures here are intended only as a guide.

randomly distributed. However, isolated point defects are not scattered at random in non-stoichiometric compounds but are often dispersed throughout the structure in some kind of regular pattern. In the next sections we try to explore the relationship between stoichiometry and structure.

It is difficult to determine the structure of compounds containing defects and it is only very recently that much of our knowledge has been formed. X-ray diffraction is the usual method for the determination of the structure of a crystal; however, this method yields the *average* structure for a crystal. For pure, relatively defect-free structures, this is a good representation but for non-stoichiometric and defect structures it avoids precisely the information that one wants to know. For this kind of structure determination a technique that is sensitive to *local* structure is needed, and such techniques are very scarce. Structures are often elucidated from a variety of sources of evidence: X-ray and neutron diffraction, density measurements, spectroscopy (when applicable) and more recently high-resolution electron microscopy (HREM); magnetic measurements have also proved useful in the case of FeO. Probably HREM has done most to clarify the understanding of defect structures, as it is capable under favourable circumstances of giving information on an atomic scale by 'direct lattice imaging'.

Non-stoichiometric compounds are of potential use to industry because their electronic, optical, magnetic and mechanical properties can be modified by changing the proportions of the atomic constituents. This is widely exploited and researched by the electronics and other industries.

3.7.2 Non-stoichiometry in wustite

Ferrous oxide is known as **wustite (FeO)**, and it has the NaCl (rock salt) crystal structure. Accurate chemical analysis shows that it is non-stoichiometric: it is always deficient in iron. The FeO phase diagram (Figure 3.17) shows that the compositional range of wustite increases with temperature and that stoichiometric FeO is not included in the range of stability. Below 570°C wustite disproportionates to α-iron and Fe_3O_4.

There are two ways in which an iron deficiency could be accommodated by a defect structure: there could either be iron vacancies giving a formula $Fe_{1-x}O$, or alternatively there could be an excess of oxygen in interstitial positions, with the formula, FeO_{1+x}. A comparison of the theoretical and measured densities of the crystal distinguishes between the alternatives. The easiest method of measuring the density of a crystal is the flotation method. Liquids of differing densities (that dissolve in one another) are mixed together until a mixture is found that will just suspend the crystal; it neither floats nor sinks. The density of that liquid mixture must then be

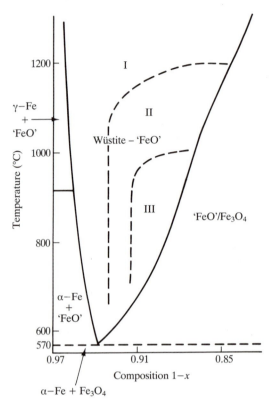

Figure 3.17 Phase diagram of the FeO system. I, II and III together comprise the wustite region.

the same as that of the crystal, and it can be found by weighing an accurately measured volume.

The *theoretical* density of a crystal can be obtained from the volume of the unit cell and the mass of the unit cell contents. The results of an X-ray diffraction structure determination gives both of these data, as the unit cell dimensions are accurately measured and the type and number of formula units in the unit cell are also determined. An example of this type of calculation for FeO follows. A particular crystal of FeO was found to have a unit cell dimension of 430.1 pm, a measured density of 5.728 kg m^{-3}, and an iron to oxygen ratio of 0.945. The unit cell volume (which is a cube) is thus $(430.1 \, \text{pm})^3 = 7.956 \times 10^7 \, (\text{pm})^3 = 7.956 \times 10^{-29} \, \text{m}^3$.

There are four formula units of FeO in a perfect unit cell with the rock-salt structure. The mass of these four units can be calculated from their relative atomic masses: Fe, 55.85 and O, 16.00. One mole of FeO weighs

$(55.85 + 16.00)\,g = 0.07185\,kg$, four moles weigh $(4 \times 0.07185)\,kg$, and four formula units weigh $(4 \times 0.07185)/N_A\,kg = 4.773 \times 10^{-25}\,kg$, where **Avogadro's number** $(N_A) = 6.022 \times 10^{23}\,mol^{-1}$.

The sample under consideration has an Fe:O ratio of 0.945. Assume, in the first instance, that it has iron vacancies. The unit cell contents in this case will be $(4 \times 0.945) = 3.78$ Fe and 4 O. The mass of the contents will be $[(3.78 \times 55.85) + (4 \times 16.00)]/(N_A \times 10^3)\,kg$. Dividing by the volume of the unit cell, we get a value of $5.742 \times 10^3\,kg\,m^{-3}$ for the density. If instead the sample possesses interstitial oxygens, the ratio of oxygens to iron in the unit cell will be given by $1/0.945 = 1.058$. The unit cell in this case will contain 4 Fe and $(4 \times 1.058) = 4.232$ O. The mass of this unit cell is given by $[(4 \times 55.85) + (4.232 \times 16.00)]/(N_A \times 10^3)\,kg$, giving a density of $6.076 \times 10^3\,kg\,m^{-3}$. Comparing the two sets of calculations with the experimentally measured density of $5.728 \times 10^3\,kg\,m^{-3}$, it is clear that this sample contains iron vacancies and that the formula should be written as $Fe_{0.945}O$. A table of densities is drawn up for FeO in Table 3.6.

Table 3.6 Experimental and theoretical densities $(10^3\,kg\,m^{-3})$ for FeO

O:Fe ratio	Fe:O ratio	Lattice parameter (pm)	Observed density	Theoretical density	
				Interstitial O	Fe vacancies
1.058	0.945	430.1	5.728	6.076	5.742
1.075	0.930	429.2	5.658	6.136	5.706
1.087	0.920	428.5	5.624	6.181	5.687
1.099	0.910	428.2	5.613	6.210	5.652

It is found to be characteristic of most non-stoichiometric compounds that their unit cell size varies smoothly with composition but the symmetry is unchanged. This is known as **Vegard's law**.

Summarizing, non-stoichiometric compounds are found to exist over a range of composition and throughout that range the unit cell length varies smoothly with no change of symmetry. It is possible to determine whether the non-stoichiometry is accommodated by vacancy or interstitial defects using density measurements.

Electronic defects in FeO

The discussion of the defects in FeO has so far been only structural. Now we turn our attention to the balancing of the charges within the crystal. In principle the compensation for the iron deficiency can be made either by oxidation of some Fe(II) ions or by reduction of some oxide anions. It is energetically more favourable to oxidize Fe(II). For each Fe^{2+} vacancy,

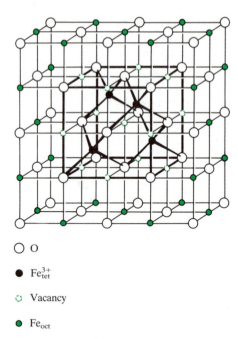

○ O

● Fe^{3+}_{tet}

○ Vacancy

● Fe_{oct}

Figure 3.18 The Koch–Cohen cluster shown with the front and back planes cut away for clarity. The central section with the four tetrahedrally coordinated Fe^{3+} ions is picked out in bold.

two Fe^{2+} cations must be oxidized to Fe^{3+}. In the overwhelming majority of cases, defect creation involves changes in the cation oxidation state. In the case of metal excess in simple compounds, we would usually expect to find that neighbouring cation(s) would be reduced.

In a later section we will look at some general cases of non-stoichiometry in simple oxides, but before we do that we will complete the FeO story with a look at its detailed structure.

The structure of FeO

FeO has the NaCl structure with Fe^{2+} ions in octahedral sites. The iron deficiency manifests itself as cation vacancies, and the electronic compensation made for this is that for every Fe^{2+} ion vacancy there are two neighbouring Fe^{3+} ions. We might reasonably expect that the Fe^{2+}, Fe^{3+}, and the cation vacancies are randomly distributed over the octahedral sites in the ccp O^{2-} array in wustite. Structural studies (X-ray, neutron and magnetic) have shown that this is not the case and that some of the Fe^{3+} ions are in *tetrahedral* sites.

Although the structure of wustite is still under debate, it appears that it contains various types of **defect cluster** and that these new structures are distributed throughout the crystal. A defect cluster is a region of the crystal where the defects form an ordered structure. One possibility is known as the **Koch–Cohen cluster** and part of it is shown in Figure 3.18. It bears a strong resemblance to the structure of Fe_3O_4, the next highest oxide of iron. We can think of wustite as fragments of Fe_3O_4 intergrown in the rock-salt structure of FeO. (The crystal structure of Fe_3O_4 is discussed in a different context in Chapter 7).

At the centre of the Koch–Cohen cluster is a modified FeO unit cell (drawn in bold in Figure 3.18). This contains four additional tetrahedrally coordinated Fe^{3+} ions in the tetrahedral holes; furthermore, the octahedrally coordinated iron sites at the centre of this unit cell, and at the midpoints of its edges, are vacant. The other octahedrally coordinated iron sites in the cluster are occupied, but they may contain either Fe(II) or Fe(III) ions. We therefore designate them simply as Fe_{oct}. The front and back planes have been cut away from the diagram in Figure 3.18 to make the central section more visible.

It is instructive to consider the composition of a cluster such as the one shown in Figure 3.18 in some detail. Clusters such as this are often referred to by the ratio of cation vacancies to interstitial Fe^{3+} in tetrahedral holes, in this case 13:4. The complete cluster, allowing for the back and front faces which are not illustrated here, contains eight NaCl-type unit cells, and thus 32 oxide ions. A cluster with no defects would also contain 32 iron cations on the octahedral sites. Taking into account the 13 octahedral vacancies, there must be 19 Fe_{oct} ions and the four interstitial Fe_{tet}^{3+} ions making 23 iron cations in all. The overall formula for the cluster is thus $Fe_{23}O_{32}$, almost Fe_3O_4!

Having determined the atomic contents of the cluster we now turn our attention to the charges. There are 32 oxide ions, so to balance them the Fe cations overall must have 64 positive charges. Twelve are accounted for by the four Fe^{3+} ions in tetrahedral positions, leaving 52 to find from the remaining 19 Fe_{oct} cations. This accounting is difficult to do by inspection. Suppose that there are xFe^{2+} ions and yFe^{3+} ions in octahedral sites, we know that:

$$x + y = 19. \tag{3.26}$$

We also know that their total charges must equal 52, so:

$$2x + 3y = 52 \tag{3.27}$$

giving two simultaneous equations. Solving gives $y = 14$ and $x = 5$.

The octahedral sites surrounding the central (bold) unit cell are thus occupied by 5 Fe^{2+} ions and by 14 Fe^{3+} ions. By injecting such clusters throughout the FeO structure, the non-stoichiometric structure is built up. The exact formula of the compound (the value of x in $Fe_{1-x}O$) will depend on the average separation of the randomly injected clusters. In the oxygen-rich limit when the whole structure is composed of these

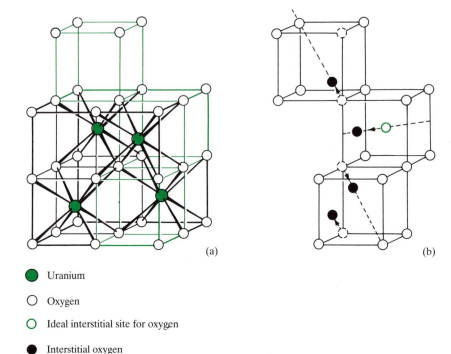

(a)   (b)

● Uranium

○ Oxygen

○ Ideal interstitial site for oxygen

● Interstitial oxygen

◌ Vacancy

Figure 3.19 (a) The fluorite structure of UO_2 with a unit cell marked in bold and the defect cluster position in colour. (b) Interstitial defect cluster in UO_{2+x}. Uranium positions (not shown) are in the centre of every other cube.

clusters there is a new ordered structure of formula $Fe_{23}O_{32}$ which is based on the structure of the parent compound. This new structure has its own unit cell, larger than that of the parent compound and of lower symmetry: the new structure is referred to as a **superstructure** or **superlattice** of the parent.

3.7.3 Uranium dioxide

Above 1127°C a single oxygen-rich non-stoichiometric phase of UO_2 is found with formula UO_{2+x}, ranging from UO_2 to $UO_{2.25}$. Unlike FeO, where a metal-deficient oxide was achieved through cation vacancies, in this example the metal deficiency arises from interstitial anions.

$UO_{2.25}$ corresponds to U_4O_9, which is a well-characterized oxide of uranium known at low temperature. UO_2 has the fluorite structure (Figure 3.19). The unit cell in Figure 3.19 contains four formula units of UO_2. (There are four uranium ions contained within the cell boundaries; the eight oxide ions come from: $(8 \times \frac{1}{8}) = 1$ at the corners, $(6 \times \frac{1}{2}) = 3$ at the face centres, $(12 \times \frac{1}{4}) = 3$ at the cell edges, and one at the cell midpoint.)

As more oxygen is taken into UO_2, the extra oxide ions go into interstitial positions. The most obvious site available is in the middle of one of the octants with no metal atom. Neutron diffraction shows that an interstitial oxide anion does not sit exactly in the centre of an octant but is displaced sideways; this has the effect of moving two other oxide ions from their lattice positions by a very small amount, leaving two vacant lattice positions. This is illustrated in Figure 3.19 where in (a) three vacant octants are picked out in colour and in (b) the positions of the one additional interstitial oxide and the two displaced oxides with their vacancies are shown. The movement of the ions from 'ideal' positions is shown by small arrows: the movement of the interstitial oxide from the centre of an octant is along the direction of a diagonal of one of the cube faces, whereas the movement of the oxide ions on lattice positions is along cube diagonals.

The atomic composition of the unit cell in Figure 3.19a when modified by the defect structure is U_4O_9; the loss of the displaced lattice oxygen at the centre of the top face is balanced by the gain of that at the centre of the bottom face. The net oxygen gain is just the single new interstitial in the central octant of Figure 3.19b. U_4O_9 is found to be the oxygen-rich limit for the UO_{2+x} non-stoichiometric structure. We can think of UO_{2+x}, therefore, as containing **microdomains** of the U_4O_9 structure within that of UO_2. The electronic compensation for the extra interstitial oxide ions will most likely be the oxidation of neighbouring U(IV) atoms to either U(V) or U(VI).

This account of the UO_2 structure is actually a slightly simplified version of the truth. There are two different (but similar) positions that the interstitial oxides can take within the structure, and when both of these are ordered throughout the structure we find a very large unit cell for U_4O_9 based on $4 \times 4 \times 4$ fluorite unit cells (with a volume of 64 times that of the UO_2 unit cell). However, the discussion above illustrates the basic principles of what is happening.

3.7.4 The titanium monoxide structure

Titanium and oxygen form non-stoichiometric phases which exist over a range of composition centred about the stoichiometric 1:1 value, from $TiO_{0.65}$ to $TiO_{1.25}$. We shall look at what happens in the upper range from $TiO_{1.00}$ to $TiO_{1.25}$.

At the stoichiometric composition of $TiO_{1.00}$ the crystal structure is based on a NaCl structure with vacancies in both the metal and the oxygen sublattices; one-sixth of the titaniums, and one-sixth of the oxygens are missing. Above 900°C, these vacancies are randomly

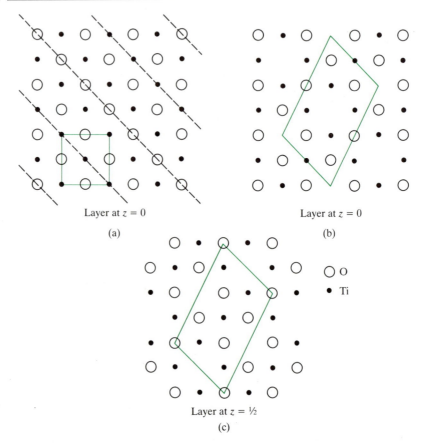

Figure 3.20 Layers parallel to the horizontal planes of Figure 1.20. (a) The hypothetical TiO structure of the NaCl type shown in Figure 1.20; the line of intersection of every third vertical diagonal plane is marked by a dashed coloured line. (b) The same plane in the observed structure of TiO; every alternate atom is removed along the diagonal lines in (a). (c) The plane directly beneath the layer in (b); again, every alternate atom is removed along the cuts made by the planes whose intersection lines are shown in (a). In (b) and (c), the cross-section of a monoclinic unit cell is indicated.

distributed but below this temperature they are ordered as shown in Figure 3.20.

In Figure 3.20a we show a layer through a NaCl-type structure. Every third vertical diagonal plane has been picked out by a dashed line. In the $TiO_{1.00}$ structure every other atom along the dashed lines is missing (Figure 3.20b). If we consider that in these first two diagrams we are looking at the structure along the y-axis, and that this layer is the top of the unit cells, $b = 0$, then the layer below this and parallel to it will be the central horizontal plane of the unit cell at $b = \frac{1}{2}$. This is drawn in Figure

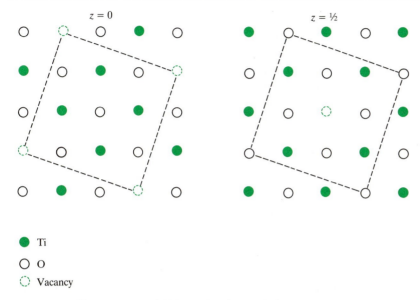

Figure 3.21 The structure of $TiO_{1.25}$ showing both O and Ti positions.

3.20c, and again we notice that every other atom along every third diagonal plane is missing. This is true throughout the structure. In the figure the unit cell of a perfect NaCl-type structure is marked on (a) whereas the boundaries of the new unit cell, taking the ordered defects into account, are marked on (b) and (c). The new unit cell of the superlattice is **monoclinic** (see Chapter 1) because one of the angles (β) is not equal to 90°. This structure is unusual in that it is stoichiometric but contains vacancies on both the anion and cation sublattices.

As noted in Chapter 2, unusually for a transition metal monoxide, $TiO_{1.00}$ shows metallic conductivity. The existence of the vacant sites within the TiO structure is thought to permit sufficient contraction of the lattice such that the $3d$ orbitals on titanium overlap, thus broadening the conduction band and allowing electronic conduction.

When titanium monoxide has the limiting formula $TiO_{1.25}$, it has a different defect structure, still based on the NaCl structure, but with all the oxygens present, and one in every five titaniums missing (Figure 3.21). The pattern of the titanium vacancies is shown in Figure 3.22 which shows a layer of the type in Figure 3.20 but with the oxygens omitted; only titaniums are marked. Drawing horizontal lines through the titaniums, will show that every fifth one is missing; again we see that the ordering of the defects has produced a superlattice. Where samples of titanium oxide have formulae which lie between the two limits discussed here, $TiO_{1.00}$ and $TiO_{1.25}$, the structure seems to consist of portions of the $TiO_{1.00}$ and

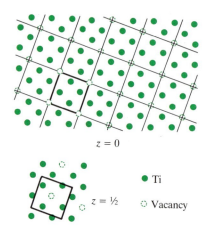

$z = 0$

$z = \frac{1}{2}$

● Ti

○ Vacancy

Figure 3.22 Successive Ti layers in the structure of $TiO_{1.25}$.

$TiO_{1.25}$ structures intergrown. Although most texts refer to the structure as $TiO_{1.25}$, as we have here, on the definitions that we have used previously when discussing 'FeO' it is more correctly written as $Ti_{0.8}O$ ($Ti_{1-x}O$), as this indicates that the structure contains titanium vacancies rather than interstitial oxygens.

3.8 PLANAR DEFECTS

In the introduction to this chapter we mentioned that crystals often contain **extended defects** as well as point defects. The simplest *linear* defect is a **dislocation** where there is a fault in the arrangement of the atoms in a line through the crystal lattice. There are many different types of *planar* defects, most of which we are not able to discuss here either for reasons of space or of complexity, such as **grain boundaries** which are of more relevance to materials scientists, and **chemical twinning** which can contain unit cells mirrored about the twin plane through the crystal. However, we shall look briefly at two forms of planar defect which are relevant to the kind of structural discussions in this book: (1) **crystallographic shear planes**, where the oxygen vacancies effectively collect together in a plane which runs through the crystal, and (2) **intergrowth structures**, where two different but related structures alternate throughout the crystal.

3.8.1 Crystallographic shear planes

Non-stoichiometric compounds are found for the higher oxides of tungsten, molybdenum and titanium, WO_{3-x}, MoO_{3-x} and TiO_{2-x}, respectively. The reaction of these systems to the presence of point defects is entirely different from what has been discussed previously; in fact the point defects are eliminated by a process known as **crystallographic shear (CS)**.

In these systems a series of closely related compounds with very similar formulae and structure exists. The formulae of these compounds all obey a general formula, which for the molybdenum and tungsten oxides can be Mo_nO_{3n-1}, Mo_nO_{3n-2}, W_nO_{3n-1} and W_nO_{3n-2} and for titanium dioxide is Ti_nO_{2n-1}; n can vary taking values of four and above. The resulting series of oxides is known as a **homologous series** (like the alkanes in organic chemistry). The first seven members of the molybdenum trioxide series are: Mo_4O_{11}, Mo_5O_{14}, Mo_6O_{17}, Mo_7O_{20}, Mo_8O_{23}, Mo_9O_{26}, $Mo_{10}O_{29}$ and $Mo_{11}O_{32}$.

In these compounds we find regions of corner-linked octahedra separated from each other by thin regions of a different structure known as the crystallographic shear planes. The different members of a homologous series are determined by the fixed spacing between the CS planes. The structure of a shear plane is quite difficult to understand and these structures are usually depicted by the linking of octahedra as described in Chapter 1.

WO_3 has several polymorphs, but above 900°C the WO_3 structure is that of ReO_3 (Figure 3.23; ignore the bold squares for the time being). (The structures of the other polymorphs are distortions of the ReO_3 structure.) ReO_3 is made up of $[ReO_6]$ octahedra that are linked together via their corners; each corner of an octahedron is shared with another. Figure 3.23a shows part of one layer of the octahedra in the structure. Notice that, within the layer, any octahedron is linked to four others; it is also linked, via its upper and lower corners, to octahedra in the layers above and below. Part of the ReO_3 structure is drawn in Figure 3.24; note that every oxygen atom is shared between two metal atoms. As six oxygens surround each Re the overall formula is ReO_3.

The non-stoichiometry in WO_{3-x} is achieved by some of the octahedra in this structure changing from corner-sharing to edge-sharing. Look back now to the octahedra marked in bold in Figure 3.23a. The edge-sharing corresponds to shearing the structure so that the chains of bold octahedra are displaced to the positions in Figure 3.23b. This shearing occurs at regular intervals in the structure and is interspersed with slabs of the 'ReO₃' structure (corner-linked $[WO_6]$ octahedra). It creates groups of four octahedra which share edges. The direction of maximum density of the edge-sharing groups is called the crystallographic shear plane and is indicated by an arrow in (b).

In order to see how crystallographic shear alters the stoichiometry of WO_3, we need to find the stoichiometry of one of the groups of four octahedra that are linked together by sharing edges; one of these groups

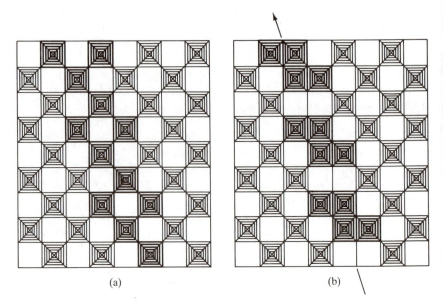

(a)   (b)

Figure 3.23 Formation of shear structure.

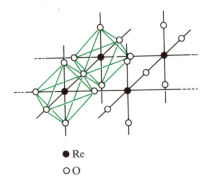

● Re

○ O

Figure 3.24 Part of the ReO_3 structure showing the linking of octahedra through the corners.

is shown in Figure 3.25. The group in this figure consists of 4 W atoms and 18 O atoms. Fourteen of the oxygen atoms are linked out to other octahedra (these bonds are indicated), so are each shared by two W atoms, while the remaining four oxygens are only involved in the edge-sharing within the group. The overall stoichiometry is given by [4 W + $(14 \times \frac{1}{2})$ O + 4 O], giving W_4O_{11}.

Clearly, if groups of four octahedra with stoichiometry W_4O_{11} are

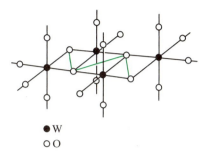

● W
○ O

Figure 3.25 Group of four [WO$_6$] octahedra sharing edges (marked in colour) formed by the creation of shear planes in W$_n$O$_{3n-1}$.

interspersed throughout a perfect WO$_3$ structure, then the amount of oxygen in the structure is reduced and we can write the formula as WO$_{3-x}$. The effect of introducing the groups of four in an ordered way can be quantified. If the structure sheared in such a way that the entire structure was composed of these groups, the formula would become W$_4$O$_{11}$. If there is one [WO$_6$] octahedron for each group of four then the overall formula becomes [W$_4$O$_{11}$ + WO$_3$] = W$_5$O$_{14}$. Clearly we can extend this process to any number of [WO$_6$] octahedra regularly interspersed between the groups:

$$W_4O_{11} + 2WO_3 = W_6O_{17}; \qquad W_4O_{11} + 3WO_3 = W_7O_{20};$$
$$W_4O_{11} + 4WO_3 = W_8O_{23}; \qquad W_4O_{11} + 5WO_3 = W_9O_{26};$$
$$W_4O_{11} + 6WO_3 = W_{10}O_{29}; \qquad W_4O_{11} + 7WO_3 = W_{11}O_{32}.$$

The basic formula of the group of four W$_4$O$_{11}$ can be written as W$_n$O$_{3n-1}$ where $n = 4$. This formula also holds for all the other formulae that are listed above. So we have produced the general formula for the homologous series simply by introducing set ratios of the edge-sharing groups in amongst the [WO$_6$] octahedra.

The shear planes are found to repeat throughout a particular structure in a regular and ordered fashion, so any particular sample of WO$_{3-x}$ will have a specific formula corresponding to one of those listed above. The different members of the homologous series are determined by the fixed spacing between the CS planes. An example of one of the structures is shown in Figure 3.26. A unit cell has been marked so that the ratio of [WO$_6$] octahedra to the groups of four is clear. Within the marked unit cell there is one group of four and seven octahedra giving the overall formula W$_4$O$_{11}$ + 7WO$_3$ = W$_{11}$O$_{32}$.

Members of the Mo$_n$O$_{3n-1}$ series have the same structure as their W$_n$O$_{3n-1}$ analogue, even though unreduced MoO$_3$ does not have the ReO$_3$ structure, but a layer structure.

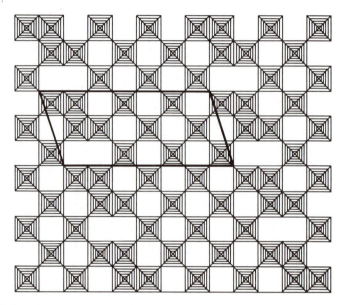

Figure 3.26 A member of the W_nO_{3n-1} homologous series with the projection of a unit cell marked.

If the structures shear in such a way that groups of six octahedra share edges regularly throughout the structure, then homologous series with the general formula M_nO_{3n-2} are formed.

The homologous series for oxygen-deficient TiO_2 is given by the formula Ti_nO_{2n-1}. In this case, the octahedra along the CS planes are joined to each other by sharing faces, whereas in the unreduced parts of the TiO_2 structure the octahedra share edges as in rutile.

3.8.2 Planar intergrowths

As this is rather a complex area we will only look at one example here, that of intergrowths in the **tungsten bronzes**. The term *bronze* is applied to metallic oxides that have a deep colour, metallic lustre and are either metallic conductors or semiconductors. The sodium-tungsten bronzes, Na_xWO_3, have colours that range from yellow to red and deep purple depending on the value of x.

We have already seen that WO_3 has the rhenium oxide (ReO_3) structure, with [WO_6] octahedra joined through the corners. This is illustrated in Figures 3.23a and 3.24. The structure contains a three-dimensional network of channels throughout the structure and it has been found that alkali metals can be incorporated into the structure in these channels. The resultant crystal structure depends on the proportion of alkali metal in the

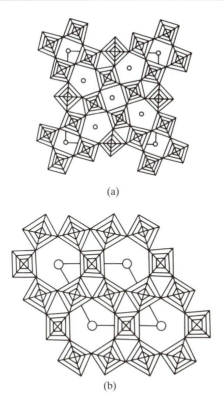

(a)

(b)

Figure 3.27 The tetragonal tungsten bronze structure (a), and the hexagonal tungsten bronze structure (b). The shaded squares represent WO_6 octahedra, which are linked to form pentagonal, square and hexagonal tunnels. These are able to contain a variable population of metal atoms, shown as open circles.

particular compound. The structures are based on three main types: *cubic* phases where the alkali metal occupies the centre of the unit cell (similar to perovskite, see Chapters 1 and 8) and *tetragonal* and *hexagonal* phases. The basic structures of two of these are illustrated in Figure 3.27. The electronic conductivity properties of the bronzes are due to the fact that charge compensation has to be made for the presence of M^+ ions in the structure. This is achieved by the change in oxidation state of some of the tungsten atoms from VI to V (such processes are discussed in more detail in section 3.10).

The hexagonal bronze structure illustrated is formed when potassium reacts with WO_3 (K needs a bigger site) and the composition lies in the range $K_{0.19}WO_3$ to $K_{0.33}WO_3$. If the proportion of potassium in the compound is less than this the structure is found to consist of WO_3 intergrown with the hexagonal structure in a regular fashion. The layers of hexagonal structure can be either one or two tunnels wide (Figure 3.28). Similar structures are observed for tungsten bronzes containing metals other than potassium, such as Rb, Cs, Ba, Sn and Pb. A high-resolution electron micrograph of the barium-tungsten bronze clearly shows its preferred single tunnel structure (Figure 3.29). In other samples the separation of the tunnels increases as the concentration of barium decreases.

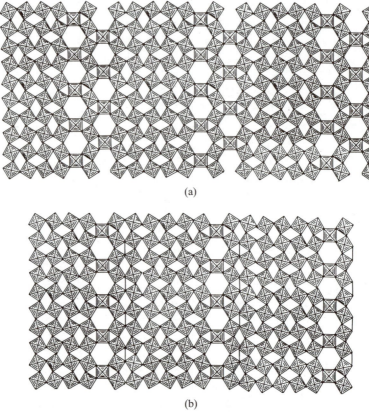

(a)

(b)

Figure 3.28 The idealized structures of two intergrowth tungsten bronze phases, (a) containing double rows of hexagonal tunnels, and (b) containing single rows of tunnels. The tungsten trioxide matrix is shown as shaded squares and the hexagonal tunnels are shown empty, although in the known intergrowth tungsten bronzes the tunnels contain variable amounts of metal atoms.

3.9 THREE-DIMENSIONAL DEFECTS

3.9.1 Block structures

In oxygen-deficient Nb_2O_5 and in mixed oxides of Nb and Ti, and Nb and W, the crystallographic shear planes occur in two sets at right angles to

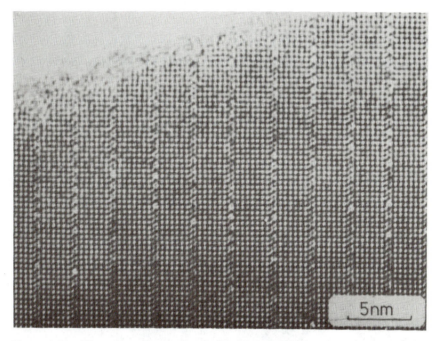

Figure 3.29 Electron micrograph of the intergrowth tungsten bronze phase Ba_xWO_3, showing the single rows of tunnels clearly. Each black spot on the image represents a tungsten atom, and many of the hexagonal tunnels seem to be empty or only partly filled with barium.

each other. The intervening regions of perfect structure now change from infinite sheets to infinite columns or blocks. These structures are known as **double shear** or **block structures** and are characterized by the cross-sectional size of the blocks. The block size is expressed as the number of octahedra sharing vertices. As well as having phases built of blocks of one size, the complexity can be increased by having blocks of two or even three different sizes arranged in an ordered fashion! The block size(s) determines the overall stoichiometry of the solid. An example of a crystal showing two different block sizes is shown in Figure 3.30.

3.9.2 Pentagonal columns

Three-dimensional faults also occur in the so-called **pentagonal column (PC) structures**. These structures contain the basic repeating unit shown in Figure 3.31a which consists of a pentagonal ring of five $[MO_6]$ octahedra. When these stack on top of one another a pentagonal column is formed that contains chains of alternating M and O atoms. These penta-

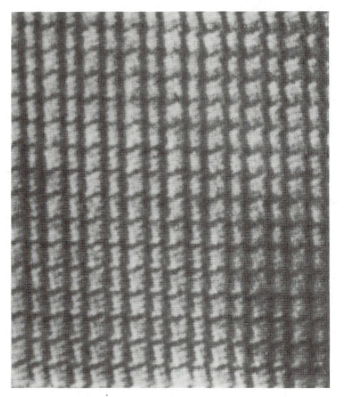

Figure 3.30 High-resolution electron micrograph of the $W_4Nb_{26}O_{77}$ structure showing strings of (4 × 4) and (3 × 4) blocks. CS planes between the blocks show as darker contrast. (Photograph by courtesy of Dr J.L. Hutchison.)

gonal columns can fit inside a ReO_3 type of structure in an ordered way, and depending on the spacing a homologous series is formed. One example is shown in Figure 3.31b for the compound Mo_5O_{14}. This type of structure is also found in the tetragonal tungsten bronzes.

3.9.3 Infinitely adaptive structures

In section 3.9.1 we saw that the mixed oxides of niobium and tungsten could have a range of different compositions made by fitting together rectangular columns or blocks. The closely related $Ta_2O_5–WO_3$ system does something even more unusual! A large number of compounds form, but they are built up by fitting together pentagonal columns (PCs). The idealized structures of two of these compounds are shown in Figure 3.32: the structures have a wave-like skeleton of PCs. As the composition varies, so the wavelength of the backbone changes, giving rise to a huge

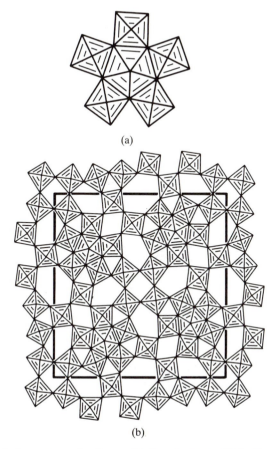

(a)

(b)

Figure 3.31   (a) A pentagonal column; (b) the structure of Mo_5O_{14}.

number of possible ordered structures, known as **infinitely adaptive compounds**.

3.10 ELECTRONIC PROPERTIES OF NON-STOICHIOMETRIC OXIDES

Earlier we considered the structure of non-stoichiometric FeO in some detail. If we apply the same principle to other binary oxides, we can define four types of compound.

Metal excess: type A – anion vacancies present, formula MO_{1-x};
                                      type B – interstitial cations, formula $M_{1+x}O$.
Metal deficiency: type C – interstitial anions, formula MO_{1+x};
                                      type D – cation vacancies, formula $M_{1-x}O$.

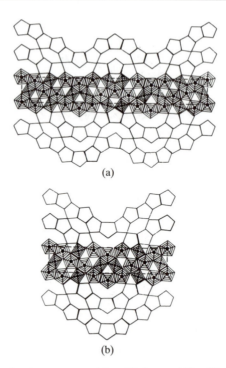

(a)

(b)

Figure 3.32 The idealized structures of $Ta_{22}W_4O_{67}$ and $Ta_{30}W_2O_{81}$. These phases are built from pentagonal columns, shown as shaded pentagons, and octahedra, shown as shaded squares. The wavelength of the chains of pentagonal columns varies with composition in such a way that *any* given anion to cation ratio can be accommodated by an ordered structure.

Figure 3.33 illustrates some of the structural possibilities for simple oxides with A, B, C and D type non-stoichiometry, assuming that they have the NaCl structure.

We looked in some detail earlier at the structure of FeO; this falls into the **type D** category. (Other compounds falling into the type D category are MnO, CoO and NiO.)

Type A oxides compensate for metal excess with **anion vacancies**. In order to maintain the overall neutrality of the crystal two electrons have to be introduced for each anion vacancy. These can be trapped at a vacant anion site (Figure 3.33a). However, it is an extremely energetic process to introduce electrons into the crystal and so we are more likely to find them associated with the metal cations (Figure 3.33b), which we can describe as reducing those cations from M^{2+} to M^+.

Type B oxides have a metal excess which is incorporated into the lattice in **interstitial** positions. This is shown in Figure 3.33a as an interstitial

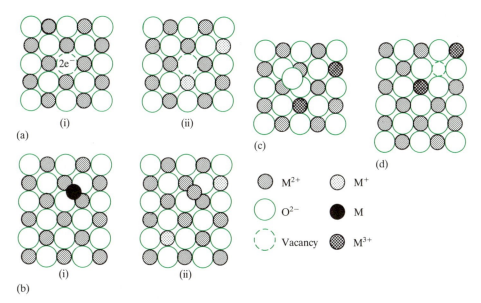

Figure 3.33 Structural possibilities for binary oxides. (a) Type A oxides: metal excess/anion vacancies. (i) This shows the two electrons that maintain charge neutrality, localized at the vacancy. (ii) The electrons are associated with the normal cations making them into M^+. (b) Type B oxides: metal excess/interstitials. (i) This shows an interstitial *atom*, whereas in (ii) the atom has ionized to M^{2+}, and the two liberated electrons are now associated with two normal cations, reducing them to M^+. (c) Type C oxides: metal deficiency/interstitial anions. The charge compensation for an interstitial anion is by way of two M^{3+} ions. (d) Type D oxides: metal deficiency/cation vacancies. The cation vacancy is compensated by two M^{3+} cations.

atom, but it is more likely that the situation in Figure 3.33b will hold, where the interstitial atom has ionized and the two electrons so released are now associated with two neighbouring ions, reducing them from M^{2+} to M^+. Cadmium oxide (CdO) has this type of structure. Zinc(II) oxide also is type B, but has the wurtzite structure.

Type C oxides compensate for the lack of metal with **interstitial anions**. The charge balance is maintained by the creation of two M^{3+} ions for each interstitial anion, each of which we can think of as M^{2+} associated with a positive hole.

Before considering the conductivity of these non-stoichiometric oxides it is probably helpful to recap what we know about the structure and properties of the stoichiometric binary oxides of the first row transition elements. A summary of the properties of binary oxides is given in Table 3.7.

Table 3.7 Properties of the first row transition element monoxides

Element	Ca	Sc	Ti	V	Cr	Mn	Fe	Co	Ni	Cu	Zn
Structure of stoichiometric oxide MO	NaCl structure	Does not exist	Defect NaCl $\frac{1}{6}$ vacancies	Defect NaCl	Does not exist	NaCl structure	(NaCl structure)*	NaCl structure	NaCl structure	PtS structure	Wurtzite structure (NaCl at high pressure)
Defect structure			$Ti_{1-\delta}O$ Ti vacancies (inter-growths of $TiO_{1.00}$ and $TiO_{1.25}$ structures)	Similar to TiO		$Mn_{1-\delta}O$ Mn vacancies	$Fe_{1-\delta}O$ Fe vacancies in defect clusters	$Co_{1-\delta}O$ Co vacancies	$Ni_{1-\delta}O$ Ni vacancies		$Zn_{1+\delta}O$ interstitial Zn
Conductivity of stoichiometric compound			Metallic	Metallic <120 K		Insulator	(Insulator)*	Insulator	Insulator		Insulator
Conductivity of non-stoichiometric compound			Metallic	Metallic		p-type hopping semi-conductor	p-type	p-type	p-type		n-type
Magnetism (Chapter 7)			Diamagnetic	Diamagnetic		Paramagnetic $\mu = 5.5\ \mu_B$ (antiferromagnetic when cooled, $T_N = 122$ K)	Paramagnetic (antiferromagnetic when cooled, $T_N = 198$ K)	Antiferromagnetic ($T_N = 292$ K)	Antiferromagnetic (paramagnetic when heated, $T_N = 530$ K)		

* Exactly stoichiometric FeO is never found.

We discussed the conductivity of the stoichiometric oxides in Chapter 2, and saw that their conductivity is dependent on two competing effects: on the one hand the d orbitals overlap to give a band – the bigger the overlap the greater the band width – and electrons in the band are delocalized over the whole structure; on the other hand, interelectronic repulsion tends to keep electrons localized on individual atoms. TiO and VO behave as metallic conductors and must therefore have good overlap of the d orbitals producing a d electron band. This overlap arises partly because Ti and V are early in the transition series (the d orbitals have not suffered the contraction, due to increased nuclear charge, seen later in the series) and partly because of the unusual crystal structure where one-sixth of the titaniums and oxygens (VO is similar) are missing from an NaCl-type structure which allows contraction of the structure and thus better d orbital overlap.

Further along the series, stoichiometric MnO, FeO, CoO and NiO are insulators. This situation is not easily described by band theory because the d orbitals are now too contracted to overlap much (typical band widths are $1\,eV$) and the overlap is not sufficient to overcome the localizing influence of interelectronic repulsions. (It is this localization of the d electrons on the atoms that gives rise to the magnetic properties of these compounds that are discussed in Chapter 7.)

Returning to the non-stoichiometric oxides, in the excess metal monoxides of types A and B, we saw that extra electrons have to compensate for the excess metal in the structure. Figure 3.33 shows that these could be associated either with an anion vacancy or alternatively they could be associated with metal cations within the structure. Although we have described this association as reducing neighbouring cations, this association can be quite weak, and these electrons can be free to move through the lattice; they are not necessarily strongly bound to particular atoms. Thermal energy is often sufficient to make these electrons move and so conductivity increases with temperature. We associate semi-conductivity with such behaviour (metallic conductivity decreases with temperature).

In Chapter 2, we discussed semiconductivity in terms of band theory. An intrinsic semiconductor has an empty conduction band lying close above the filled valence band. Electrons can be promoted into this conduction band by heating, leaving positive holes in the valence band. The current is carried by both the electrons in the conduction band and by the positive holes in the valence band. Semiconductors such as silicon can also be doped with impurities to enhance their conductivity. For instance, if a small amount of phosphorus is incorporated into the lattice the extra electrons form impurity levels near the empty conduction band and are easily excited into it. The current is now carried by the electrons in the conduction band and the semiconductor is known as **n-type** (**n** for negative). Correspondingly, doping with gallium increases the conductivity by creating positive holes in the valence band and such semiconductors are called **p-type** (**p** for positive).

Compounds of type A and B would produce *n*-type semiconductors because the conduction is produced by electrons. Conduction in these non-stoichiometric oxides is not easily described by band theory, for the reasons given earlier for their stoichiometric counterparts: the inter-electronic repulsions have localized the electrons on the atoms. So it is easier to think of the conduction electrons (or holes) localized or trapped at atoms or defects in the crystal rather than delocalized in bands throughout the solid. Conduction then occurs by jumping or **hopping** from one site to another under the influence of an electric field. In a perfect ionic crystal where all the cations are in the same valence state this would be an extremely energetic process. However, when two valence states, such as Ni^{2+} and Ni^{3+}, are available, as in these transition metal non-stoichiometric compounds, the electron jump between them does not take much energy. Although we cannot develop this theory here, we can note that the conduction in these so-called **hopping semiconductors** can be described by the equations of diffusion theory in much the same way as we did earlier for ionic conduction. We find that the mobility of a charge carrier (either an electron or a positive hole), μ, is an activated process and we can write:

$$\mu \propto \exp(-E_a/kT), \tag{3.28}$$

where E_a is the activation energy of the hop, and is of the order of 0.1–0.5 eV. The hopping conductivity is given by the expression:

$$\sigma = ne\mu, \tag{3.29}$$

where n is the number of mobile charge carriers per unit volume and e is the electronic charge. (Notice that these equations are analogous to equations (3.8) and (3.7) in Section 3.3, describing ionic mobility, μ, and ionic conductivity, σ, respectively). The density of mobile carriers, n, depends only on the composition of the crystal, and does not vary with temperature. From equation (3.28) we can see that, as for ionic conductivity, the hopping electronic conductivity increases with temperature.

In the type C and D monoxides we have shown the lack of metal as being compensated by oxidation of neighbouring cations to M^{3+}. The M^{3+} ions can be regarded as M^{2+} ions associated with a positive hole. Accordingly, if sufficient energy is available conduction can be thought to occur via the positive hole hopping to another M^{2+} ion and the electronic conductivity in these compounds will be *p*-type. MnO, CoO, NiO and FeO are materials that behave in this way. This behaviour of hopping semiconduction was described for NiO in Chapter 2 in terms of electron hopping. Regarding the charge carriers as positive holes is simply a matter of convenience and the description of a positive hole moving from Ni^{3+} to Ni^{2+} is the same as saying that an electron moves from Ni^{2+} to Ni^{3+}.

Non-stoichiometric materials can be listed which cover the whole range of electrical activity from metal to insulator. Here we have considered some metallic examples which can be described by band theory (TiO, VO) and others (such as MnO) which are better described as hopping semiconductors. Other cases, such as WO_3 and TiO_2, fall in between these extremes and a different description again is needed. Non-stoichiometric compounds such as calcia-stabilized zirconia and β-alumina are also good ionic conductors. Indeed, stabilized zirconia exhibits both electronic and ionic conductivity though, fortunately for its industrial usefulness, electronic conduction only occurs at low oxygen pressures. It is thus difficult to make generalizations about this complex behaviour and each case is best treated individually.

Semiconductor properties are extremely important to the modern electronics industry, which is constantly searching for new and improved materials. Much of their research is directed at extending the composition range and thus the properties of these materials. The composition range of a non-stoichiometric compound is often quite narrow (a so-called line phase), so to extend it (and thus extend the range of its properties also) the compound is doped with an impurity. To take one example: if we add Li_2O to NiO and then heat to high temperatures in the presence of oxygen, Li^+ ions become incorporated in the lattice and the resulting black material has the formula $Li_xNi_{1-x}O$, where x lies in the range 0–0.1. The equation for the reaction (using stoichiometric NiO for simplicity) is given by:

$$\tfrac{1}{2}xLi_2O + (1-x)NiO + \tfrac{1}{4}xO_2 = Li_xNi_{1-x}O. \tag{3.30}$$

To compensate for the presence of the Li^+ ions, Ni^{2+} ions will be oxidized to Ni^{3+} or the equivalent of a high concentration of positive holes located at Ni cations.

This process of creating electronic defects is called **valence induction** and it increases the composition range of 'NiO' tremendously. Indeed, at high Li concentrations the conductivity approaches that of a metal (although it still exhibits semiconductor behaviour in that its conductivity increases with temperature).

3.11 CONCLUSIONS

In this chapter we have tried to give some idea of the size and complexity of this subject and also its fascination, without it becoming overwhelming.

The main point to emerge from our explorations is that the concept of random, isolated point defects does not explain the complex structures of non-stoichiometric compounds but that there are many different ways for defects to become either ordered or even eliminated.

FURTHER READING

Defects

General

Tilley, R.J.D. (1987) *Defect Crystal Chemistry*, Blackie, London.
 A very readable book, written for materials scientists, but with much material relevant to this book. Suitable for higher level undergraduates and postgraduates.
Greenwood, N.N. (1968) *Ionic Crystals, Lattice Defects and Nonstoichiometry*, Butterworths, London.
 Detailed coverage of crystal structures and non-stoichiometry. Now out of print.
Cox, P.A. (1987) *The Electronic Structure and Chemistry of Solids*, Chapter 7, Oxford University Press, Oxford.
 See comments at the end of Chapter 2.
West, A.R. (1984) *Solid State Chemistry and its Applications*, Chapters 9 and 13, John Wiley, New York.
 See comments at the end of Chapter 2.
Adams, D.M. (1974) *Inorganic Solids*, Chapter 9, John Wiley, New York.
 See comments at the end of Chapter 1.

Planar defects

Mandelcorn, L. (ed.) (1963) *Nonstoichiometric Compounds*, article by A.D. Wadsley, Academic Press, New York.

Infinitely adaptive structures

Anderson, J.S. (1973) J.C.S. Dalton, *Journal of the Chemical Society*, 1107.

Photography

Hannay, N.B. (1976) *Treatise on Solid State Chemistry*, Vol. 4, Chapter 7, Plenum, New York.
 Advanced seven volume series on the solid state written by specialists in the field.

Colour centres

Nassau, K. (1983) *The Physics and Chemistry of Color*, Chapter 9, John Wiley, New York.
 Useful sections on F-centres, luminescence, phosphors and lasers.

Sodium-sulphur battery

Sudworth, J.L. and Tilley, A.R. (1985) *The Sodium Sulfur Battery*, Chapman & Hall, London.
Very specialized text containing much industrial detail.

QUESTIONS

1. If $\Delta H_m^{\ominus}$ for the formation of Schottky defects in a certain MX crystal is $200\,kJ\,mol^{-1}$, calculate n_S/N and n_S per mole for the temperatures 300, 500, 700 and 900 K.

2. Table 3.8 gives the variation of defect concentration with temperature for CsI. Determine the enthalpy of formation for one Schottky defect in this crystal.

Table 3.8 Defect concentration data for CsI

$T(K)$	n_S/N
300	1.08×10^{-16}
400	1.06×10^{-12}
500	2.63×10^{-10}
600	1.04×10^{-8}
700	1.43×10^{-7}
900	4.76×10^{-6}

3. If we increase the quantity of impurity (say $CaCl_2$) in the NaCl crystal, how will this affect the plot shown in Figure 3.8? How is the transition point in the graph affected by the purity of the crystal?

4. In NaCl, the cations are more mobile than the anions. What effect, if any, do you expect small amounts of the following impurities to have on the conductivity of NaCl crystals: (a) AgCl; (b) $MgCl_2$; (c) NaBr; (d) Na_2O?

5. Unlike fluorides, pure oxides with the fluorite structure (MO_2) show high anion conduction only at elevated temperatures, above about 2300 K. Suggest a reason for this.

6. Figure 3.34 shows the fluorite structure with the tetrahedral environment of one of the anions depicted in colour. (The anion behind this tetrahedron has been omitted for clarity.) Suppose the coloured anion jumps to the octahedral hole at the body centre. Describe, and sketch, the pathway it takes in terms of the changing coordination by cations.

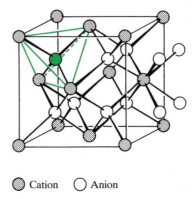

Cation   Anion

Figure 3.34 The fluorite structure, showing the coordination tetrahedron (colour) around one of the anions (also colour).

7. Make a simple estimate of the energy of defect formation in the fluorite structure. (a) Describe the coordination by nearest neighbours and next-nearest neighbours of an anion both for a normal lattice site and for an interstitial site at the centre of the unit cell shown in Figure 3.3a. (b) Use the potential energy of two ions, given by:

$$E = -\frac{e^2 Z}{4\pi\varepsilon_0 r}$$
$$= -(2.31 \times 10^{-28}\,\mathrm{J\,m})\, Z/r,$$

where Z is the charge on the other ion and r is the distance between them, to estimate the energy of defect formation in fluorite; $a = 537\,\mathrm{pm}$.

8. The zinc blende structure of γ-AgI, a low-temperature polymorph of AgI, is shown in Figure 3.35. Discuss the similarities and differences between this structure and that of α-AgI. Why do you think that the conductivity of the Ag^+ ions is lower in γ-AgI?

9. The compounds in Table 3.4 mostly contain either I^- ions or ions from the heavier end of Group 6. Explain.

10. Undoped β-alumina shows a maximum conductivity and minimum activation energy when the sodium excess is around 20–30 mole %. Thereafter, further increase in the sodium content causes the conductivity to decrease. By contrast, β-alumina crystals doped with Mg^{2+} have a much higher conductivity than do undoped crystals. Explain these observations.

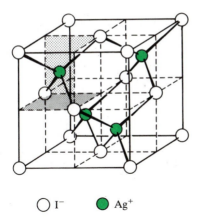

$\bigcirc$ I⁻ $\bullet$ Ag⁺

Figure 3.35 The zinc blende structure of γ-AgI.

11. Confirm the presence of iron vacancies for a sample of wustite which has a unit cell dimension of 428.2 pm, an Fe:O ratio of 0.910 and an experimental density of $6.210 \times 10^3 \, kg \, m^{-3}$.

12. How does the change in lattice parameter of 'FeO' with iron content corroborate the iron vacancy model and refute an oxide interstitial model?

13. How would you expect the formation of colour centres to affect the density of the crystal?

14. Figure 3.36 shows the central section of a possible defect cluster for FeO. (a) Determine the vacancy:interstitial ratio for this cluster. (b) Assuming that this section is surrounded by Fe ions and oxide ions in octahedral sites as in the Koch–Cohen cluster, determine the formula of a sample made totally of such clusters. (c) Determine the numbers of Fe^{2+} and Fe^{3+} ions in octahedral sites.

15. Use Figure 3.20 to confirm that TiO is a one-sixth defective NaCl structure, by counting up the atoms in the monoclinic cell.

16. Figure 3.21 shows layers in the $TiO_{1.25}$ structure with both the Ti and O sites marked. Use this and Figure 3.22 to demonstrate that the unit cell shown has the correct stoichiometry for the crystal.

17. How would you expect charge neutrality to be maintained in $TiO_{1.25}$?

18. Take a simple case where two metal oxide octahedra wish to eliminate oxygen by sharing. How does the formula change as they (i) share a corner, (ii) share an edge, (iii) share a face?

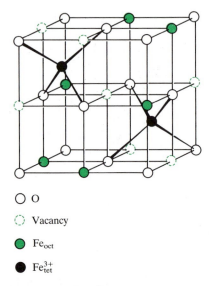

○ O

⊙ Vacancy

● Fe_{oct}

● Fe_{tet}^{3+}

Figure 3.36 A possible cluster in $Fe_{1-x}O$.

Figure 3.37 A member of the W_nO_{3n-1} homologous series.

19. Figure 3.37 shows a member of the homologous series, W_nO_{3n-1}. What formula does it correspond to?

20. ZnO is a type B (excess metal) material. What do you expect to happen to its electronic properties if it is doped with Ga_2O_3 under reducing conditions?

ANSWERS

1. $n_S \approx N \exp^{-\Delta H_S/2RT}$, where $\Delta H_S = 200 \, kJ \, mol^{-1}$ and $R = 8.314 \, J \, K^{-1} \, mol^{-1}$.

Table 3.9 Schottky defect concentration in MX compound at various temperatures

Temperature (°C)	Temperature (K)	n_S/N	$n_S/(mol^{-1})$
27	300	3.87×10^{-18}	2.33×10^6
127	500	3.57×10^{-11}	2.15×10^{13}
427	700	3.45×10^{-8}	2.08×10^{16}
627	900	1.57×10^{-6}	9.45×10^{17}

2. Rearrange equation (3.4) and take logs:

$$\frac{n_S}{N} \approx \exp^{-\Delta H_S/2RT}$$

$$\ln \frac{n_S}{N} = -\frac{\Delta H_S}{2RT},$$

or

$$\log \frac{n_S}{N} = -\frac{\Delta H_S}{2.303 \times 2RT}.$$

A plot of $\log n_S/N$ against $1/T$ gives a straight line plot passing through the origin. The slope of this graph, $-[(\Delta H_S)/(2.303 \times 2R)]$, gives a value of $183.4 \, kJ \, mol^{-1}$ for ΔH_S. Dividing by Avogadro's number, 6.022×10^{23}, gives the enthalpy of formation of one Schottky defect, $3.045 \times 10^{-19} \, J$.

3. Increasing the impurity levels does not affect the intrinsic (left-hand side) of the graph. It does, however, increase the value of σ (and thus of $\ln \sigma T$) in the extrinsic region. As the activation energy, E_a, for cation movement stays the same, the slope of the graph is unchanged (Figure 3.38). The purer the crystal, the lower the transition temperature at which thermally generated defects take over.

4. (a) Almost none. (b) Conductivity increases as two Na^+ ions are replaced by each Mg^{2+} ion, thus creating a vacancy each time. (c) Almost none. (d) Two Cl^- ions are replaced by each O^{2-} ion, creating anion vacancies; this will result in a small increase in conductivity because, although the major charge carrier is Na^+, some current can be carried by the chloride ions.

5. The oxide has more highly charged ions (M^{4+} and O^{2-}) than a fluoride. This means that the coulombic interactions will be stronger

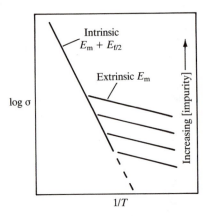

Figure 3.38 Schematic ionic conductivity of doped NaCl crystals. Parallel lines in the extrinsic region correspond to different dopant concentrations.

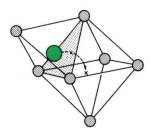

Figure 3.39.

and thus the energy of defect formation will be higher (see Table 3.1), so reducing the defect concentration at a given temperature.

6. The path is shown in Figure 3.39. The anion jumps from the tetrahedral site, through a trigonal position on the common triangular face and into the octahedral site at the body centre.

7. The anion at the body centre of the unit cell in Figure 3.3c is surrounded by six anions at distance $a/2$ and by four cations at a distance of $0.43a$. The interstitial site at the body centre of the unit cell in Figure 3.3a is surrounded by eight anions at a distance of $0.43a$ and by six cations at a distance of $a/2$.

For the normal anion site:

$r = 0.43 \times 537 \times 10^{-12}$ m and $Z = +2$, for interaction with four cations. $r = 0.5 \times 537 \times 10^{-12}$ m and $Z = -1$, for interaction with the six anions.

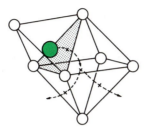

Figure 3.40.

$$E = -\left(\frac{2.31 \times 10^{-28}\,\mathrm{J\,m}}{537 \times 10^{-12}\,\mathrm{m}}\right)\left(\frac{4 \times 2}{0.43} + \frac{6 \times (-1)}{0.5}\right)$$
$$= -(4.302 \times 10^{-19}\,\mathrm{J})(6.605)$$
$$= -2.84 \times 10^{-18}\,\mathrm{J}.$$

For the interstitial site:

$$E = -(4.302 \times 10^{-19}\,\mathrm{J})\left(\frac{8 \times (-1)}{0.43} + \frac{6 \times 2}{0.5}\right)$$
$$= -(4.302 \times 10^{-19}\,\mathrm{J})(5.395)$$
$$= -2.32 \times 10^{-18}\,\mathrm{J}.$$

The energy of defect formation is the difference in energy between the two sites and so is given by:

$$(-2.32 \times 10^{-18}\,\mathrm{J}) - (-2.84 \times 10^{-18}\,\mathrm{J}) = 5.2 \times 10^{-19}\,\mathrm{J}.$$

The experimental value for fluorite is given in Table 3.1 as $4.49 \times 10^{-19}\,\mathrm{J}$. This calculation gives a very good level of agreement considering that we have ignored all the more distant interactions, internuclear repulsion, lattice vibrations and lattice relaxation!

8. Both structures contain Ag^+ ions in tetrahedral sites, but in γ-AgI half of the tetrahedral holes are occupied, so there is only one vacant equivalent site per Ag^+ ion, whereas in α-AgI there are five. Both structures contain vacant octahedral and trigonal sites. In γ-AgI the octahedral holes lie at the body centre and at the midpoint of each edge. The trigonal sites lie on the triangular faces of the octahedra, where adjacent octahedra and tetrahedra join (Figure 3.40). Both structures have the same monovalent ions and a polarizable anion.

The jump from one tetrahedral site to another in γ-AgI could take either of the routes sketched out in Figure 3.40, which first pass through a trigonal face and then through an octahedral hole. Alternatively (not shown) if the ion passed through only the top half of the octahedron it would be five-coordinate before passing through another three-coordinate face. All these routes would be of higher energy than the pathways we looked at for α-AgI.

9. I^-, S^{2-}, Se^{2-} and Te^{2-} are all polarizable anions.

10. In undoped β-alumina, the excess Na^+ ions are balanced by additional oxygens in the conduction plane. Beyond a certain point, it seems likely that these will begin to block the motion of the Na^+ ions, as observed. By contrast, doping with Mg^{2+} suggests an alternative charge compensation mechanism: simple substitution for Al^{3+} ions in the spinel-like blocks allows extra Na^+ ions into the conduction planes without the need for oxygen interstitials. This is what happens in practice.

11. Unit cell volume is $(428.2\,\text{pm})^3 = 7.8513 \times 10^{-29}\,\text{m}^3$. Mass of contents for iron vacancies:

$[(4 \times 55.86 \times 0.910) + (4 \times 16.00)]/(N_A \times 10^3)\,\text{kg}$ giving a density of $5.6534 \times 10^{-3}\,\text{kg}\,\text{m}^{-3}$.

Mass of contents for oxygen interstitial:

$[(4 \times 55.85) + (4 \times 16.00 \times 1/0.910)]/(N_A \times 10^3)\,\text{kg}$ giving a density of $6.2126 \times 10^3\,\text{kg}\,\text{m}^{-3}$.

Comparing these theoretical values with the experimental value, we again see that the evidence supports an iron vacancy model.

12. From Table 3.6, we see that as the Fe:O ratio decreases the unit cell volume also decreases; this is the trend we would expect to see as more vacancies are introduced. If the interstitial model were correct, as the Fe:O ratio decreases, the number of interstitial oxygens rises and we would expect to see a slight increase in lattice parameter.

13. A crystal containing F-centres contains anion vacancies. We would expect, therefore, that the density would be lower than that of the colourless crystal.

14. (a) The central section has two Fe^{3+} ions in tetrahedral sites and seven vacancies, so it is known as a 7:2 cluster. (b) There are 32 oxide anions. There are seven octahedral vacancies and two Fe^{3+}_{tet} interstitial ions, so there will be a total of 27 Fe cations. (There are two Fe^{3+} ions and six Fe_{oct} ions enclosed within the cluster. The outer layer will be the same as the Koch–Cohen cluster with $(8 \times \frac{1}{8})$ = $1\,Fe_{oct}$ at the corners; $(12 \times \frac{1}{4}) = 3\,Fe_{oct}$ at the midpoints of the edges; and $(30 \times \frac{1}{2}) = 15\,Fe_{oct}$ on the faces; this makes 27 Fe ions in total.) The formula would be $Fe_{27}O_{32}$. (c) The 32 oxide ions provide 64 negative charges to be balanced. The two tetrahedral Fe^{3+} ions reduce this to 58 to be balanced by the ions in octahedral positions.

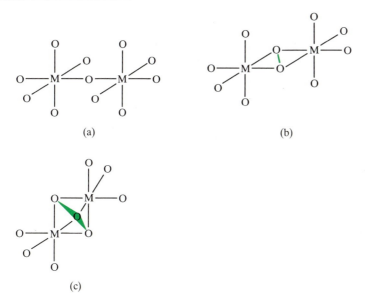

Figure 3.41 (a) Sharing a corner M_2O_{11}. (b) Sharing an edge M_2O_{10}. (c) Sharing a face M_2O_9.

Setting up simultaneous equations, we know that if x is the number of Fe^{2+} ions and y the number of Fe^{3+}, then:

$$x + y = 25$$

and adding up the charges we get:

$$2x + 3y = 58.$$

Solving gives $x = 17$ and $y = 8$.

15. **Titanium vacancies**: there are eight at the corners $(8 \times \frac{1}{8}) = 1$, and $(2 \times \frac{1}{2}) = 1$ on cell faces. **Titanium ions**: cell edges, $(4 \times \frac{1}{4}) = 1$. Cell faces, $(8 \times \frac{1}{2}) = 4$ on the top and bottom. There are five ions contained within the cell boundary, making 10 in total. The titanium stoichiometry of the unit cell is obviously representative of the whole structure: of the 12 sites, 10 are occupied and two are vacant. This is also true for oxygen. **Oxygen vacancies**: cell faces, $(4 \times \frac{1}{2}) = 2$. **Oxide ions**: cell faces, $(8 \times \frac{1}{2}) = 4$. Cell edges, $(8 \times \frac{1}{4}) = 2$. There are four ions contained within the cell boundary, making 10 in all.

16. Taking the titanium positions first: there are vacancies at all eight corners $(8 \times \frac{1}{8}) = 1$ and one vacancy in the centre of the cell, making two vacancies in all. There are no Ti^{2+} on cell edges. There are Ti^{2+} on only two of the faces, the top and bottom, which have four each, contributing $(2 \times 4 \times \frac{1}{2}) = 4$. There are 4 Ti^{2+} enclosed within the

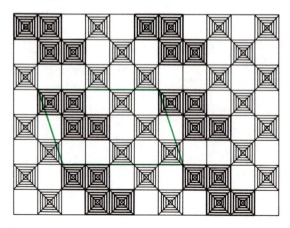

Figure 3.42 The structure of W_8O_{23} with the projection of a unit cell marked in colour.

cell. Four edges have O^{2-} ions giving $(4 \times 1 \times \frac{1}{4}) = 1$; only the top and bottom faces have O^{2-} ions, and these have five each contributing $(2 \times 5 \times \frac{1}{2}) = 5$; there are 4 O^{2-} ions enclosed within the cell, making 10 in all. The overall content of this unit cell is thus, Ti_8O_{10}, which corresponds to $TiO_{1.25}$.

17. The structure has Ti vacancies, every fifth Ti is missing. For every absent Ti^{2+} ion there must be two Ti^{3+} present or one Ti^{4+}.

18. The effects of sharing are shown in Figure 3.41.

19. Figure 3.42 shows the shear structure with a unit cell added. Within the boundary there is one group of four edge-sharing octahedra and 4 $[WO_6]$ octahedra. The formula is thus $W_4O_{11} + 4WO_3 = W_8O_{23}$.

20. Non-stoichiometric ZnO is an *n*-type semiconductor. Gallium ions entering the structure of ZnO have a charge of +3. If the Ga^{3+} substitutes for Zn^{2+} and the crystal maintains its stoichiometry, oxygen will be lost during the reaction. The electrons made available from the oxide ions becoming oxygen molecules will remain in the structure to effect the necessary charge compensation, thus enhancing the *n*-type semiconduction. An equation for the reaction is:

$$xGa_2O_3 + 2(1-x)ZnO = 2Ga_xZn_{1-x}O + \tfrac{1}{2}xO_2,$$

where for simplicity we have used 1:1 ZnO.

Low-dimensional solids | 4

4.1 INTRODUCTION

There are a number of solids whose properties are not isotropic. They may, for example, conduct electrons along planes, but not between planes, or they may only conduct along chains of atoms. Such solids are known as **low-dimensional solids**. There has been recent interest in developing devices using such materials, for example in batteries, in solar cells and in photocopiers, and there has been theoretical interest because they can be treated as if they were one- or two-dimensional, and the equations for one and two dimensions are simpler than for three dimensions. Examples of these solids are found in both organic and inorganic chemistry. Much research interest has centred on organic compounds; a polymer with low density and metallic conductivity would be extremely useful technologically, for example, in switching devices, sensors, 'smart' windows. The first low-dimensional metals found were, however, inorganic. Examples are given of both organic and inorganic solids, starting with one-dimensional electronic conductors.

4.2 ONE-DIMENSIONAL SOLIDS

In Chapter 2, we took some hypothetical one-dimensional solids – a chain of hydrogen atoms and a chain of lithium atoms. The solids studied below contain chains of atoms or molecules which can be treated in a similar manner to the hypothetical chain. In the crystals of these solids the chains are not completely isolated, but because the forces between chains are much less than those bonding the chains, the chains can be treated as one-dimensional solids. The first solid we consider is an organic polymer, polyacetylene.

4.2.1 Polyacetylene

Most organic polymers are electrical insulators; a mouldable plastic that was a conductor would be a technological breakthrough. To be

Figure 4.1 *cis*- and *trans*-polyacetylene.

conducting, the polymer must have electrons delocalized along the chain length. In small conjugated alkenes such as butadiene with alternate double and single bonds, the π electrons are delocalized over the molecule. If we take a very long conjugated olefin, we might expect to obtain a band of π levels, and if this band were partly occupied, we would expect to have a one-dimensional conductor. Polyacetylene is just such a conjugated long chain polymer. This is formed by polymerizing ethyne (acetylene) and has two forms, *cis* and *trans* (Figure 4.1).

If polyacetylene consisted of a regular evenly spaced chain of carbon atoms, the highest occupied energy band, the π band, would be half full and polyacetylene would be an electrical conductor. In practice, polyacetylene shows only modest electrical conductivity, comparable with semiconductors such as silicon: the *cis* form has a conductivity of the order of $10^{-7} \, S \, m^{-1}$ and the *trans* form, $10^{-3} \, S \, m^{-1}$. The crystal structure is difficult to determine accurately, but diffraction measurements indicate that there is an alternation in bond lengths of about 6 pm. This is much less than would be expected for truly alternating single and double bonds (C—C, 154 pm in ethane; C=C, 134 pm in ethene). Nonetheless this does indicate that the electrons are tending to localize in double bonds rather than be equally distributed over the whole chain. What in fact is happening is that two bands, bonding and antibonding, are forming with a band gap where non-bonding levels would be expected. There are just enough electrons to fill the lower band. This leads to a lower energy than the half-full single band (Figure 4.2). This splitting of the band is an example of Peierls' theorem which asserts that a one-dimensional metal is always electronically unstable with respect to a non-metallic state; there is always some way of opening an energy gap and creating a semiconductor.

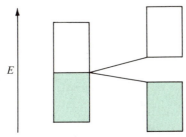

Figure 4.2 The band gap in polyacetylene produced by the alternation of long and short bonds along the chain.

Figure 4.3 A film of polyacetylene forms on the inner surface of the reaction vessel, after ethyne gas passes over the catalyst solution on the walls. The paper-thin flexible sheet of polyacetylene is then stripped from the walls prior to doping.

Thus, although a very simple bonding picture of such solids would suggest a half-filled band and metallic conductivity, the best that can be expected is a semiconducting polymer. Furthermore, there were other problems in finding useful conducting polymers. At first, attempts to make polyacetylene and similar solids resulted in short chain molecules or amorphous, unmeltable powders. Then in 1961, Hatano and co-workers in Tokyo managed to produce thin films of polyacetylene, and 10 years later Shirakawa and Ikeda made films of *cis*-polyacetylene which could be converted into the *trans* form. They achieved this by directing a stream of ethyne gas on to the surface of a Ziegler–Natta catalyst (a mixture of triethyl aluminium and titanium tetrabutoxide). To make a large film, the catalyst solution can be spread in a thin layer over the walls of a reaction vessel (Figure 4.3), and then ethyne gas allowed to enter. The polyacetylene produced in this way has a smooth shiny surface on one side and a sponge-like structure. It can be converted to the thermodynamically more stable *trans* form by heating. The conversion

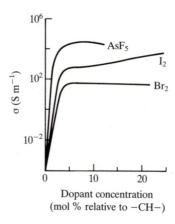

Figure 4.4 Plot showing the increase in conductivity of polyacetylene on the addition of various dopants.

is quite rapid above 370 K. After conversion, the smooth side of the film is silvery in appearance, becoming blue when the film is very thin.

A method to improve the conductivity was discovered when Shirakawa visited McDiarmid and Heeger in Pennsylvania later in the 1970s. The Americans had been working on smaller conjugated molecules to which they added an electron acceptor in order to make them conducting. It was a natural step to try this approach with polyacetylene. If an electron acceptor such as bromine is added to polyacetylene, it takes electrons from the lower π bonding band, forming $[(CH)^{\delta+}Br_{\delta}^{-}]_n$. The doped polyacetylene now has holes in its valence band and, like p-type semiconductors, its conductivity is greater than that of the undoped material. Other examples of dopants that can oxidize polyacetylene are I_2, AsF_5 and $HClO_4^{-}$. The effect of these dopants is shown in Figure 4.4, where it can be seen that the conductivity rises from $10^{-3}\,S\,m^{-1}$ to as much as $10^5\,S\,m^{-1}$ using only small quantities of dopant.

The conductivity of polyacetylene is also increased by dopants that are electron donors. For example, the polymer can be doped with alkali metals to give, for example, $[Li_{\delta}^{+}(CH)^{\delta-}]_n$. The electron donor dopants add electrons to the upper π band, making this partly full, and so producing an n-type semiconductor. The wide range of conductivities produced by these two forms of doping is illustrated in Figure 4.5.

Conducting polyacetylene has a variety of possible uses but no practical devices have appeared so far, primarily because of its susceptibility to attack by the oxygen in the atmosphere. The polymer loses its metallic lustre and becomes brittle when exposed to air. Suggested uses include a lightweight rechargeable battery and a photovoltaic cell. In the electrochemical cell proposed as the basis for the battery, two electrodes made

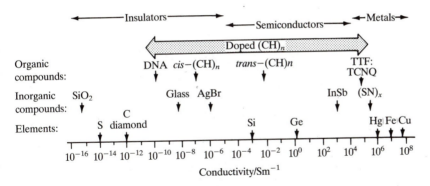

Figure 4.5 The conductivity of undoped and doped polyacetylene ($(CH)_n$) compared with the values for some of the better-known insulators, semiconductors and metals.

of polyacetylene film are dipped into a solution of a lithium salt such as $LiAsF_6$ or $LiBF_4$ in propylene carbonate. When a direct current is passed through the cell, the polyacetylene at the anode is oxidized and that at the cathode reduced. The electrode reactions are

$$(CH)_n + xnBF_4^- = [(CH)^{x+}(BF_4^-)_x]_n + xne^- \qquad (4.1)$$

and

$$(CH)_n + xnLi^+ + xne^- = [Li_x^+(CH)^{x-}]_n. \qquad (4.2)$$

The $p-n$ junction needed for a photovoltaic cell, as described in Chapter 2, can be made by bringing together a p-type film and an n-type film of polyacetylene.

4.2.2 Platinum chain compounds

Platinum chain compounds were the first one-dimensional electronic conductors found. Crystals containing chains of tetracyanoplatinate and bisoxalatoplatinate ions were prepared in the nineteenth century. The compound $K_2Pt(CN)_4 \cdot 3H_2O$ is an insulator, but when it is dissolved in water and oxidized with a little bromine, copper-coloured needles of $[K_2Pt(CN)_4]Br_{0.3} \cdot 3H_2O$ can be crystallized out. This salt is commonly known as KCP or KCP(Br). In the late 1960s, K. Krogmann noticed that such salts had unusual conductivities and these compounds are often named after him. Such salts are unlikely to be of great commercial interest due to the weight and price of platinum. Their unusual electrical and optical properties have, however, intrigued chemists ever since their discovery.

The KCP crystals contain chains of closely spaced square-planar $[Pt(CN)_4]^{2-}$ units (Figure 4.6) and the conductivity parallel to the chains

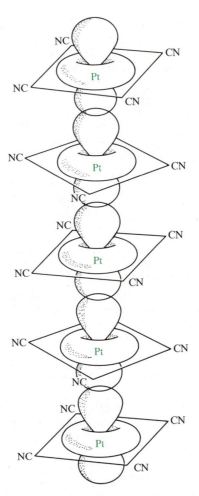

Figure 4.6 Columnar stacked structure of $[Pt(CN)_4]^{x-}$ ions.

is of the order of 10^4 times that perpendicular to the chains. The optical properties also resemble those of a one-dimensional metal in that light polarized parallel to the chains is reflected, giving the copper colour and metallic sheen, whereas the crystals are transparent to visible light polarized perpendicular to the chain.

The Pt–Pt separation is 289 pm along the chain; this is quite close to the Pt–Pt distance found in Pt metal (277 pm). The ligands surrounding the metal atoms ensure that the interchain separation is large (about 800 pm) so that there is no overlap of Pt orbitals at right angles to the stacking direction. The separation of the chains can be seen in the packing diagram (Figure 4.7). The bromide ions are shown in the centre

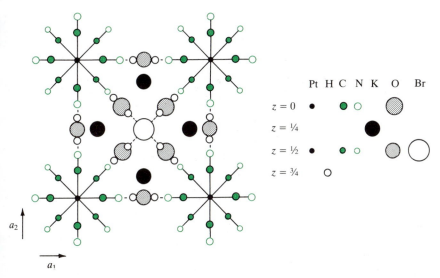

	Pt	H	C	N	K	O	Br
$z = 0$	●		●	○		⬤ (shaded)	
$z = \frac{1}{4}$						● (black)	
$z = \frac{1}{2}$	●		●	○		⬤ (shaded)	○
$z = \frac{3}{4}$	○						

Figure 4.7 Crystal structure of KCP, showing a projection along the conducting c-axis. You are looking down a Pt chain at each corner of the unit cell: the staggering of the CN ligands can be seen.

of the unit cell. These occupy only 60% of such sites and are arranged randomly among them. Water molecules and potassium ions also occupy space between the chains.

This arrangement gives us a clue to the origin of the properties of these salts. The $5d$ orbitals on the Pt atoms in a chain can overlap, but there is no overlap of $5d$ orbitals on Pt atoms in different chains. If we take the chain axis to be the z-axis, then we can think of all the Pt atoms in one chain being bound together by electrons occupying orbitals formed from overlapping d_{z^2} orbitals. The overlapping $5d$ orbitals are shown in Figure 4.6. Since there are a large number of atoms in a chain, the situation is very like the chain of lithium atoms discussed in Chapter 2 and a band of orbitals delocalized over the chain is formed.

In the parent compound, $K_2Pt(CN)_4 \cdot 3H_2O$, there are just enough electrons to fill this band and so the compound is an insulator. When bromine is added, electrons are taken from the top of this band to form bromide ions. There is now a partially filled band and so metallic conductivity would be expected. A closer look at the properties of KCP shows that the explanation of the conductivity is not quite that simple. KCP is a metallic conductor at room temperature, but as the temperature is lowered below 150 K, the conductivity is found to drop sharply. The conductivity of metals increases as the temperature is lowered, and so it appears that a band gap is forming at low temperatures. The explanation lies with Peierls' theorem, that suggests that a one-dimensional conductor

Figure 4.8 Periodic lattice distortion showing modulation of regular chain spacing. The vertical lines indicate the changes in bond length (exaggerated).

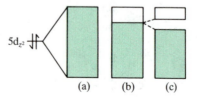

$5d_{z^2}$

(a)   (b)   (c)

Figure 4.9 Simple band structure of a one-dimensional metallic complex. (a) Filled d band; (b) partially filled d band; (c) splitting of a partially filled band by Peierls distortion.

cannot be formed; there will always be a structural change that will lead to a filled band and a band gap.

The way that Peierls' theorem operates in Pt chain complexes is shown by a careful X-ray diffraction study of another Pt chain compound, $Rb_{1.67}[Pt(C_2O_4)_2] \cdot \frac{3}{2}H_2O$. (This compound also has electrons missing from the $5d$ orbitals but in this case they are compensated for by a deficiency of cations.) This crystal contains chains of Pt atoms, but the Pt–Pt separations are not all the same. In fact, three distinct Pt–Pt distances are found: 272, 283 and 302 pm; these repeat regularly throughout the crystal. In Figure 4.8, this is illustrated by showing the Pt atoms by black dots and then adding vertical lines between two atoms to represent the bond lengths (all the bond distances are exaggerated for clarity). A line drawn through the tops of all the vertical lines show a wave-like pattern rather like a sine wave. So the Pt–Pt bond distances vary periodically along the chain. As a consequence, the electron density will vary periodically along the chain; the density will be higher between atoms that are closer together and the bonding between these atoms will be stronger. This regular variation of electron density is called a **charge density wave**. The variation in bond length is the distortion that produces a band gap (Figure 4.9), this periodically varying chain being of lower electronic energy than a regular chain and thus satisfying Peierls' theorem.

The X-ray study was done on $Rb_{1.67}[Pt(C_2O_4)_2] \cdot \frac{3}{2}H_2O$ because the bond length differences are fairly large in this salt, and because the variation in bond length coincided with the crystal lattice in a simple fashion so that it could be measured by traditional diffraction methods.

Table 4.1 Platinum chain compounds and their properties

Compound	Abbreviation	Pt–Pt bond length (pm)	Conductivity $(S\,m^{-1})$
$K_2[Pt(CN)_4]\,Br_{0.3} \cdot 3H_2O$	KCP(Br)	289	30×10^3
$Rb_2[Pt(CN)_4](FHF)_{0.4}$	RbCP(FHF)	280	230×10^3
$Cs_2[Pt(CN)_4]\,Cl_{0.3}$	CsCP(Cl)	286	20×10^3
$K_{1.75}[Pt(CN)_4] \cdot \frac{3}{2}H_2O$	K(def)CP	296	$0.5{-}10 \times 10^3$
$Rb_{1.67}[Pt(C_2O_4)_2] \cdot \frac{3}{2}H_2O$	Rb–OP	$\left\{ \begin{array}{l} 272 \\ 283 \\ 302 \end{array} \right.$	0.7
$Mg_{0.82}[Pt(C_2O_4)_2] \cdot 6H_2O$	Mg–OP	285	$0.02{-}5 \times 10^3$

A periodic variation of Pt–Pt distances also appears in KCP but this is smaller. At low temperatures, then, these Pt chain compounds will distort to produce a band gap and thus will be semiconductors. At room temperature, the electronic levels interact with lattice vibrations and the variation in bond length disappears.

Many Pt chain compounds have been made in addition to KCP. Some, like the rubidium salt mentioned above, are cation-deficient. Others have chloride or even bifluoride ions (HF_2^-) in place of bromide. Table 4.1 lists a selection of these salts and their properties. Note that the ligands attached to Pt are always composed of small atoms, typically C, N, O. Ligands with larger atoms such as S or P would not allow the Pt atoms to approach sufficiently closely to form chains. Although most work has been done on Pt, similar chain compounds are also found with other metals, particularly those with d^8 configurations which favour square planar complexes, for example iridium in oxidation state 1.

4.2.3 Other one-dimensional solids and molecular metals

Apart from polyacetylene, there are other examples of conjugated, conducting polymers: polypyrrole, polythiophene, polyaniline and polyphenylenevinylene (Figure 4.10). The description of the bonding and conductivity of these will be essentially similar to that of polyacetylene.

There is, however, another class of organic one-dimensional conductors in which there are chains composed of stacks of two unsaturated cyclic molecules arranged alternately. The first one of this class discovered was TTF–TCNQ. The two molecules involved are tetrathiafulvalene (TTF) and tetracyanoquinonedimethane (TCNQ). These are shown in Figure 4.11, which also shows how they are stacked in the crystal.

The electrical conductivity of TTF–TCNQ is of the order of $10^2\,S\,m^{-1}$

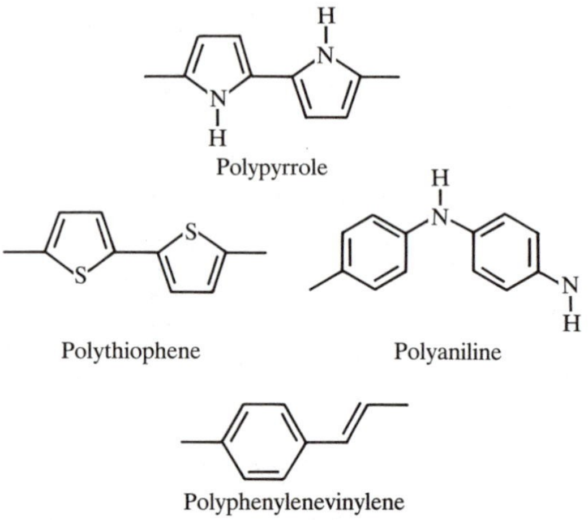

Polypyrrole

Polythiophene

Polyaniline

Polyphenylenevinylene

Figure 4.10 Repeating units of some conducting polymers.

at room temperature and increases with decreasing temperature until around 80 K when the conductivity, like that of KCP, drops as the temperature is lowered. TCNQ is a good electron acceptor and will, for example, accept electrons from alkali metal atoms to form ionic salts. In TTF–TCNQ, the columns of each type of molecule interact to form delocalized orbitals. Some electrons from the highest energy filled band of TTF move across to partly fill a band of TCNQ, so that both types of columns have partially occupied bands. The number of electrons transferred corresponds to about 0.69 electrons per molecule. This partial transfer only occurs with molecules such as TTF whose electron donor ability is neither too small nor too large. With poor electron donors, no charge transfer occurs. With very good electron donors such as alkali metals one electron per TCNQ is transferred and the acceptor band is full. Thus $K^+(TCNQ)^-$ is an insulator. At low temperatures, TTF–TCNQ suffers a periodic distortion similar to that in KCP and so its conductivity drops.

Molecular solids such as TTF–TCNQ which have metallic conductivity are often referred to as **molecular metals** or synthetic metals. Poly-acetylene, KCP and TTF–TCNQ are all described as one-dimensional electronic conductors, because there is little interaction between chains in the crystal. There is, however, another class of molecular metals which while appearing to resemble these solids, is less unambiguously defined as one-dimensional. For example, tetramethyl-tetraselenofulvalene (TMTSF; Figure 4.12) forms a series of salts with inorganic anions. The crystals of these salts contain stacks of TMTSF molecules and the TMTSF molecules carry a fractional charge (0.5+). As expected, these salts have high electronic conductivities at room temperatures. Unlike

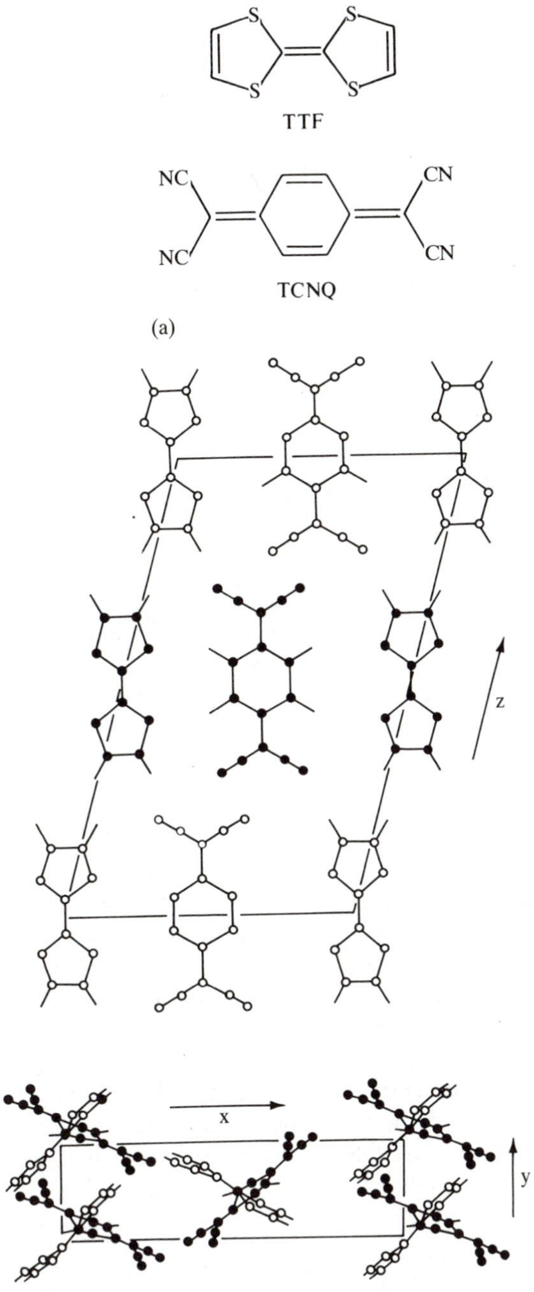

(a)

(b)

Figure 4.11 Structures of (a) TTF and TCNQ and (b) solid TTF–TCNQ showing alternate stacks of TTF and TCNQ molecules.

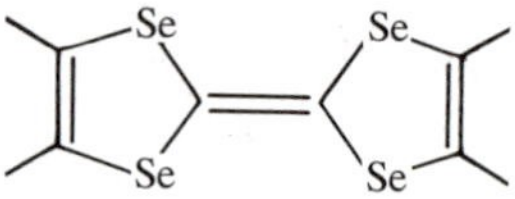

Figure 4.12 Structure of tetramethyl-tetraselenofulvalene, TMTSF.

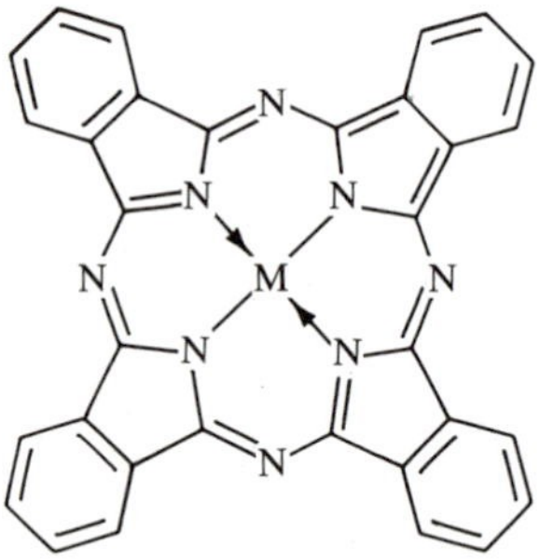

Figure 4.13 Phthalocyanine macrocycle coordinated to a metal M. The macrocycle can have hydrogen atoms or other groups attached to it.

TTF–TCNQ, however, $(TMTSF)_2^+(ClO_4)^-$ and other similar salts remain highly conducting at low temperatures and indeed at very low temperatures become superconducting. The reason for this seems to be that there is significant overlap between stacks so that the one-dimensional model is not as valid as for, say, TTF–TCNQ, and, in particular, Peierls' theorem no longer holds. Other materials of this type with significant interaction between chains include $(SN)_x$ and $Hg_{3-x}AsF_6$, both of which become superconducting at low temperatures.

An interesting class of molecular metals are the **metallophthalocyanines**. The basic building block of these is shown in Figure 4.13. These form stacks and there are parallel stacks of counterions. When partially oxidized, the crystals of these compounds have very high electrical conductivities in the direction of the stacks which persist down to very low temperatures (1.5 K). Here the metals interact as in KCP, but the conduction electrons are also delocalized over the large unsaturated ligands and are not confined to the metal ions.

4.3 TWO-DIMENSIONAL SOLIDS

Many solids adopt layer structures, particularly metal salts with less electronegative anions such as sulphide or iodide or with metals to

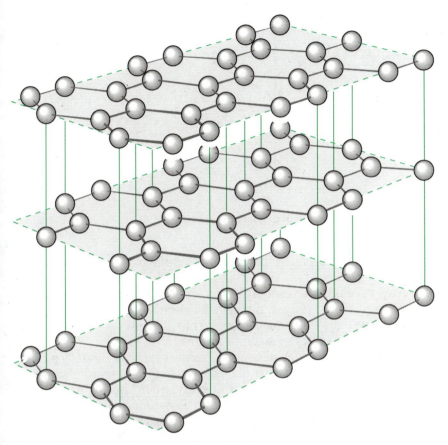

Figure 4.14 Hexagonal structure of graphite layers.

the bottom right of the periodic table such as lead or cadmium. Such structures show that a purely ionic model is not appropriate for these solids and that there is considerable covalent or metallic bonding present. We shall be concerned here with those solids for which the layer structure produces interesting properties, particularly electronic properties. We consider an example of a layered metal sulphide which has been the subject of much research due to its potential use in batteries, but first this section takes a rather different example of a two-dimensional conductor.

4.3.1 Graphite

Graphite is a very familiar substance, with many uses. The lead in lead pencils is graphite, and finely divided forms of graphite are used to absorb gases and solutes. Its absorption properties find a wide range of applications from gas masks to decolouring food. Graphite formed by

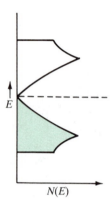

Figure 4.15 Band structure of graphite.

pyrolysis of oriented organic polymer fibres is the basis of carbon fibres. Graphite is also used as a support for several industrially important catalysts. Its electronic conductivity is exploited in several industrial electrolysis processes where it is used as an electrode. Crystals of graphite are, however, only good conductors in two dimensions, and it is this two-dimensional aspect that we are concerned with here.

Crystals of graphite contain layers of interlocking hexagons (Chapter 1; Figure 4.14). Figure 4.14 shows the most stable form in which the layers are stacked ABAB with the atoms in every other layer directly above each other. There is another form which also contains layers of inter-locking hexagons but in which the layers are stacked ABCABC. Each carbon can be thought of as having three single bonds to neighbouring carbons. This leaves one valence electron per carbon in a p orbital at right angles to the plane of the layer. These p orbitals combine to form delocalized orbitals which extend over the whole layer. If the layer contains n carbon atoms, then n orbitals are formed, and there are n electrons to fit in them. Thus half of the delocalized orbitals are filled. Were the orbitals to form one band, then this would explain the con-ductivity of graphite since there would be a half filled band confined to the layers. The situation is close to this, but not quite as simple. The delocalized orbitals in fact form two bands, one bonding and one anti-bonding. (This is reminiscent of diamond, where as shown in Chapter 2, the s/p band splits with a gap where non-bonding orbitals would lie.) The lower band is full and the upper band empty. Graphite is a conductor because the band gap is zero, and electrons are thus readily promoted to the upper band. The band structure for graphite is shown in Figure 4.15. Because the density of states is low at the Fermi level, the conductivity is not as high as that for a typical metal. It can, however, be increased as shown below.

4.3.2 Intercalation compounds of graphite

Because the bonding between layers in graphite is weak, it is easy to insert molecules or ions into the spaces between layers. The solids produced by reversible insertion of such guest molecules into lattices are known as **intercalation compounds**, and although originally applied to layered solids this term is now taken to include other solids with similar host–guest interactions. Since the 1960s attention has been paid to intercalation compounds as being of possible importance as catalysts and as electrodes for high-energy density batteries.

Many layered solids form intercalation compounds, but graphite is particularly interesting because it forms compounds with both electron donors and electron acceptors. Amongst electron donors, the most extensively studied are the alkali metals. The alkali metals enter graphite between the layers and produce strongly coloured solids in which the layers of carbon atoms have moved further apart. For example, potassium forms a golden compound KC_8 in which the interlayer spacing is increased by 200 pm. The potassium donates an electron to the graphite (forming K^+) and the conductivity of the graphite now increases because it has a partially full antibonding band.

The first graphite intercalation compound was made in 1841. This contained sulphate, an electron acceptor. Since then many other electron acceptor intercalation compounds have been made with, for example, NO_3^-, CrO_3, Br_2, $FeCl_3$ and AsF_5. In these compounds, the graphite layers donate electrons to the inserted molecules or ions, thus producing a partially filled bonding band. This increases the conductivity and some of these compounds have electrical conductivity approaching that of aluminium.

In graphite, the current is carried through the layers by delocalized p electrons. There are very few examples of other layered solids with delocalized p electrons, but there are a number of transition metal salts with d electrons delocalized over layers, and we now discuss one of these.

4.3.3 Titanium disulphide: the lithium–titanium disulphide battery

Titanium disulphide (TiS_2) has a CdI_2 structure, as mentioned in Chapter 2. This structure was discussed in Chapter 1 (see Figure 1.31) and a view of TiS_2 is illustrated in Figure 4.16. The solid is golden-yellow and has a high electrical conductivity along the titanium layers. As shown in Chapter 2, the conductivity of TiS_2 arises from a very small band gap or from a slight overlap of the $S3p$ valence band and $Ti3d$ conduction band. This situation is similar to that in graphite. The band structure is repeated here (Figure 4.17).

As in graphite, the conductivity of TiS_2 can be increased by forming

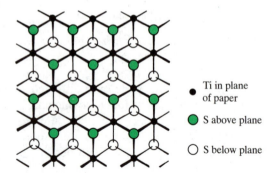

Figure 4.16 The structure of TiS$_2$ viewed from above.

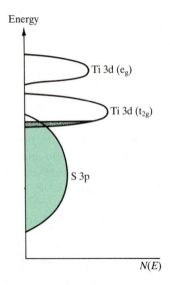

Figure 4.17 Band structure of TiS$_2$.

intercalation compounds. These are generally formed with molecules that are electron donors, however, in contrast to the intercalation compounds of graphite which are formed with both electron donors and electron acceptors. The electron donors can be metals such as the alkali metals or copper, or organic molecules such as amines. The electrons from the donor go into the conduction band of TiS$_2$, increasing the number of electrons able to carry current.

One possible use of the intercalation compounds is in rechargeable batteries and Whittingham in the 1970s developed a battery that operated at room temperature based on the intercalation of Li in TiS$_2$. In the **lithium–titanium disulphide battery**, one electrode is lithium metal and

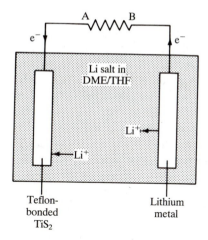

Figure 4.18 The Li–TiS$_2$ battery during a discharge phase.

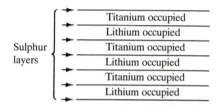

Figure 4.19 Occupation pattern of octahedral holes between close-packed layers of sulphur atoms in LiTiS$_2$.

the other is TiS$_2$ bonded to a polymer such as teflon. The electrolyte is a lithium salt dissolved in an organic solvent. Typically, the solvent is a mixture of dimethoxyethane (DME) and tetrahydrofuran (THF). The set up is shown schematically in Figure 4.18. When the circuit is complete, lithium metal from the lithium electrode dissolves giving solvated ions, and solvated ions in the solution are deposited in the TiS$_2$. These ions intercalate into the disulphide, and the charge is balanced by electrons from the external circuit. Thus the two electrode reactions are:

$$Li(s) = Li^+(solv) + e^- \tag{4.3}$$

and

$$xLi^+(solv) + TiS_2(s) + xe^- = Li_xTiS_2(s), \tag{4.4}$$

giving an overall reaction:

$$xLi(s) + TiS_2(s) = Li_xTiS_2(s). \tag{4.5}$$

The intercalation compound formed has layers of a variable number of lithium ions between sulphide layers (Figure 4.19).

When the TiS_2 has completely discharged, the battery can be recharged by applying a voltage across AB. A voltage high enough to overcome the free energy of the cell reaction will cause lithium ions to return to the solvent and lithium to be deposited on the metal electrode, thus restoring the battery to its original condition. The usefulness of the $Li-TiS_2$ electrode depends on the electrode material being a good conductor and on the electrode reaction being reversible. Li_xTiS_2 is a particularly good electrical conductor, because not only is the electronic conductivity increased by the donation of electrons to the conduction band, but also the lithium ions themselves act as current carriers and so the compound is both an electronic and an ionic conductor. (Ionic conductors are described in Chapter 3.)

Titanium disulphide is probably the best-known example of a layered transition metal sulphide, but layered structures are also found for the disulphides of Zr, Hf, V, Nb, Ta, Mo and W. At one time there was considerable interest in these compounds because some of their inter-calates were found to be superconductors at very low temperatures. It was hoped that by altering the interlayer spacing, a compound would be found that was superconducting at higher temperatures. Unfortunately, it soon transpired that altering the spacing by inserting different mole-cules had very little effect on the temperature at which superconductivity appeared. It was concluded that the superconductivity was confined to the layers. When high temperature superconductors were discovered, it was in a different class of compound as will be described in Chapter 8.

FURTHER READING

Hoffmann, R. (1988) *Solids and Surfaces*, VCH Publishers, Inc, New York.
  An unusual solid state chemistry book. The approach is probably more suitable for postgraduates and it concentrates on surfaces, but has sections on one- and two-dimensional solids.
Cox, P.A. (1987) *Electronic Structure and Chemistry of Solids*, Oxford University Press, Oxford.
  See comments following Chapter 2. This contains a section on low-dimensional conductors.

  The following two articles at an approachable level were written by re-searchers in one dimensional metallic complexes.
Day, P. (1983) Low-dimensional solids, *Chemistry in Britain*, April, 306.
Underhill, A.E. and Watkins, D.M. (1980) One-dimensional metallic complexes, *Chemical Society Reviews*, 429.

Bloor, D. (1983) Organic conductors, *Chemistry in Britain*, September, 725.
This article discusses low-dimensional organic conductors such as polyacetylene.

QUESTIONS

1. Which orbitals would you expect to combine to form a delocalized band in polyphenylenevinylene?

2. Would doping of polyacetylene with (a) Rb, (b) H_2SO_4 give an *n*-type or a *p*-type conductor?

3. When polyacetylene $(CH)_n$ is doped with chloric(VII) acid, $HClO_4$, some of the acid is used to oxidize the polyacetylene and some to provide a counter anion. The oxidation reaction for the acid is given in equation (4.6). Write a balanced equation for the overall reaction.

$$ClO_4^- + 8H^+ + 8e^- = Cl^- + 4H_2O. \tag{4.6}$$

4. What is the oxidation state of platinum in KCP?

5. $(SN)_x$ remains a metallic conductor down to very low temperatures. Why does this suggest that there is interaction between the chains in this solid?

6. In Chapter 2, it was shown that VO_2 is only metallic above 340 K. Below this temperature, VO_2 adopts a crystal structure in which cation–cation distances in one direction are alternately 265 and 312 pm. By analogy with polyacetylene, suggest why this would lead to the solid being non-magnetic and semiconducting.

7. HMTTF–TCNQ is metallic but HMTTF–TCNQF$_4$ is not. Suggest a possible explanation for the difference. HMTTF = hexamethylenetetrathiafulvalene, TCNQ = tetracyanoquinodimethane, and TCNQF$_4$ = tetracyanotetrafluoroquinodimethane.

ANSWERS

1. Polyphenylenevinylene has a π system delocalized over the benzene ring. It is likely that this is also delocalized over the conjugated double bond and hence to the benzene ring of the next unit. Thus like polyacetylene, this polymer will have a delocalized π system, but it will include the π ring orbitals.

2. (a) Rb would lose an electron to the polyacetylene, forming Rb^+, and give n-type polymer with a partially full conduction band. (b) H_2SO_4 would act as an electron acceptor and give p-type polyacetylene.

3. $$8(CH)_n + 9\delta n HClO_4 = 8[(CH)^{\delta+}(ClO_4)_\delta^-]_n + \delta n HCl + 4\delta n H_2O.$$
$$(4.7)$$

4. The potassium ions have oxidation state $+1$, the cyanide ions will contribute -1 and the bromine ions -1 also. Since the net oxidation state is 0, that of platinum is $-2 + 4 + 0.3 = 2.3$. Note that oxidation state is a formal concept and that the above result does not imply that there is a charge of 2.3 on each platinum.

5. If there were no interaction between the chains, $(SN)_x$ would be a one-dimensional conductor and it would be expected at low temperatures to obey Peierls' theorem. This would result in an insulating or semi-conducting solid. The observation of low temperature conductivity suggests that Peierls' theorem is not obeyed and one ready explanation of this is that the chains interact so that the solid no longer behaves like a one-dimensional system.

6. The alternating bond length suggests a partial localization of the d electrons. In polyacetylene, such bond alternation is the means by which a band gap is opened up and the same is probably true here. Thus instead of the partly full band expected if the oxide were to adopt the TiO_2 structure, the d band splits so that there is a full band and an empty band.

7. Fluorine tends to attract electrons to itself and so $TCNQF_4$ is likely to be a stronger electron acceptor than TCNQ. The conducting properties of solids like TTF–TCNQ arise from partial transfer of electrons from one type of molecule to the other. $TCNQF_4$ is a sufficiently good acceptor to enable transfer of one electron per unit, thus completely filling the conduction band in one stack and completely emptying it in the other stack. So HMTTF–$TCNQF_4$ has no partially full bands and hence is not a metallic conductor.

Zeolites

5.1 INTRODUCTION

For many years zeolites have been very useful as cation exchangers, as used in water softening, and also as molecular sieves for separating molecules of different sizes and shapes (e.g. as drying agents). More recently, however, research has focused on their ability to act as catalysts in a wide variety of reactions, many of them highly specific, and they are now used extensively in industry for this purpose. Approximately 40 naturally occurring zeolites have been characterized, but in the quest for new catalysts more than 130 entirely synthetic structures have been developed.

Zeolites were first described as a mineral group by the Swedish mineralogist Baron Axel Cronstedt in 1756. They are a class of crystalline **aluminosilicates** based on rigid anionic frameworks with well-defined channels and cavities. These cavities contain exchangeable metal cations (Na^+, K^+, etc.) and can also hold removable and replaceable guest molecules (water in naturally occurring zeolites). It is their ability to lose water on heating that earned them their name; Cronstedt observed that on heating with a blowtorch they hissed and bubbled as though they were boiling and named them zeolites from the Greek words *zeo*, to boil and *lithos*, stone.

5.2 COMPOSITION AND STRUCTURE

The general formula for the composition of a zeolite is:

$$M_{x/n}[(AlO_2)_x(SiO_2)_y] \cdot mH_2O,$$

where cations M of valence n neutralize the negative charges on the aluminosilicate framework.

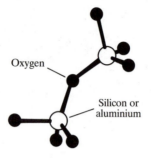

Figure 5.1 The zeolite building units. Two SiO_4/AlO_4 tetrahedra linked by corner-sharing.

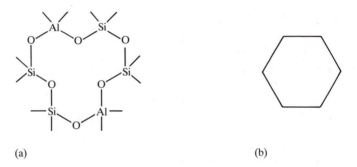

(a)   (b)

Figure 5.2 (a) 6-ring containing two Al and four Si atoms. (b) Shorthand version of the same 6-ring.

5.2.1 Frameworks

The primary building units of zeolites are $[SiO_4]^{4-}$ and $[AlO_4]^{5-}$ tetrahedra (see Chapter 1, section 1.5.7) linked together by **corner-sharing**, forming bent oxygen bridges (Figure 5.1). Silicon–oxygen tetrahedra are electrically neutral when connected together in a three-dimensional network as in quartz, SiO_2. The substitution of Si(IV) by Al(III) in such a structure, however, creates an electrical imbalance, and to preserve overall electrical neutrality, each $[AlO_4]$ tetrahedron needs a balancing positive charge. This is provided by exchangeable cations held electrostatically within the zeolite.

It is possible for the tetrahedra to link by sharing two, three, or all four corners, thus forming a variety of different structures. The linked tetrahedra are usually illustrated by drawing a straight line to represent the oxygen bridge connecting two tetrahedral units. In this way the six linked tetrahedra in Figure 5.2a are simply represented by a hexagon

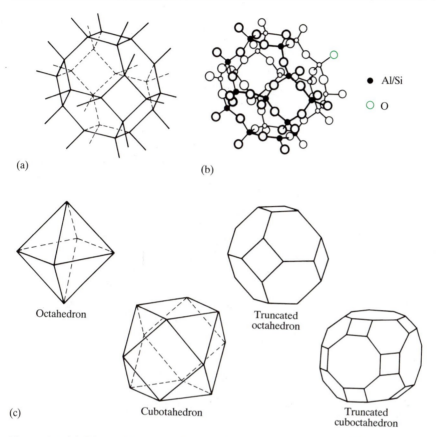

Al/Si ●
O ○

(a)

(b)

Octahedron

Truncated octahedron

(c)   Cubotahedron

Truncated cuboctahedron

Figure 5.3 (a) The sodalite unit showing the atomic positions. (b) The truncated octahedron of linked tetrahedra which form the sodalite unit. (c) The relationship between an octahedron, a truncated octahedron, a cuboctahedron and a truncated cuboctahedron.

(Figure 5.2b). This is known as a **6-ring**, and a tetrahedrally coordinated atom occurs at each intersection between two straight lines. (Remember, however, that the oxygen bridge is not itself linear.)

Many zeolite structures are based on a secondary building unit that consists of 24 silica or alumina tetrahedra linked together; here we find 4- and 6-rings linked together to form a basket-like structure called a **truncated octahedron**. This is the **sodalite unit** (or **β-cage**). Figure 5.3a shows the linked tetrahedra of the sodalite unit including the oxygen bridges, and the simplified schematic diagram is shown in Figure 5.3b. Notice that in this three-dimensional structure a tetrahedral atom is located at the intersection of four lines, as oxygen bridges are made by corner-sharing from all four vertices of the tetrahedron.

Several of the most important zeolite structures are based on the

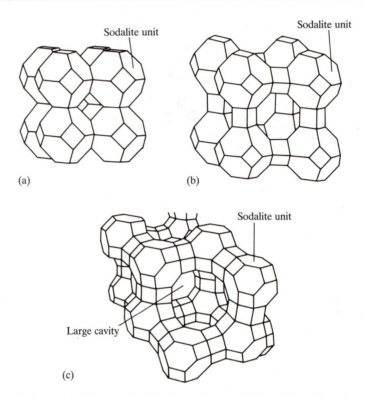

Sodalite unit

Sodalite unit

(a)   (b)

Sodalite unit

Large cavity

(c)

Figure 5.4 Zeolite frameworks built up from sodalite units: (a) sodalite, (b) zeolite-A, (c) faujasite (zeolites-X and -Y).

sodalite unit (Figure 5.4). The mineral **sodalite** itself is composed of these units, with each 4-ring shared by two β-cages in a primitive array. Notice that the cavity enclosed by the eight sodalite units shown in Figure 5.4a is itself a sodalite unit (i.e. sodalite units are space-filling). This is a highly symmetrical structure and the cavities are linked together to form channels which run parallel to all three cubic crystal axes.

A synthetic zeolite, **zeolite-A** (also called Linde A) is shown in Figure 5.4b. Here the sodalite units are again stacked in a primitive array, but now they are linked by oxygen bridges between the 4-rings. A three-dimensional network of linked cavities forming channels is thus formed. The formula of zeolite-A is given by:

$$Na_{12}[(SiO_2)_{12}(AlO_2)_{12}] \cdot 27H_2O.$$

In this fairly typical example, the Si/Al ratio is unity, and we find that in the crystal structure the Si and Al atoms strictly alternate.

The structure of **faujasite**, a naturally occurring mineral, is shown in

Figure 5.4c. The sodalite units are linked by oxygen bridges between four of the eight 6-rings in a tetrahedral array, forming hexagonal prisms. The synthetic zeolites, **zeolite-X** and **zeolite-Y** (Linde X and Linde Y) also have this basic underlying structure. Zeolite-X structures have a Si/Al ratio between 1 and 1.5, whereas zeolite-Y structures have Si/Al ratios between 1.5 and 3. The tetrahedral array encloses a large **supercage** (also known as an **α-cage**) which is entered through a 12-ring window.

5.2.2 Si/Al ratios

We saw that zeolite-A has a Si/Al ratio of 1. Some zeolites have quite high Si/Al ratios: zeolite **ZK-4**, with the same framework structure as zeolite-A, has a ratio of 2.5. Many of the new synthetic zeolites that have been developed for catalysis are highly siliceous: **ZSM-5** (**Z**eolite **S**ocony-**M**obil) can have a Si/Al ratio which lies between 20 and ∞ (the latter being virtually pure SiO_2); this far outstrips the ratio of 5.5 found in mordenite, which is the most siliceous of the naturally occurring zeolite minerals.

Clearly, changing the Si/Al ratio of a zeolite also changes its cation content; the fewer aluminium atoms there are, the fewer exchangeable cations will be present. The highly siliceous zeolites are inherently hydrophobic in character, and their affinity is for hydrocarbons.

5.2.3 Exchangeable cations

The zeolite Si/Al–O framework is rigid, but the cations are not an integral part of this framework and are often called **exchangeable cations**: they are fairly mobile and readily replaced by other cations (whence their use as cation exchange materials).

The presence and position of the cations in zeolites is important for several reasons. The cross-section of the rings and channels in the structures can be altered by changing the size or the charge (and thus the number) of the cations, and this significantly affects the size of the molecules that can be adsorbed. A change in cationic occupation also changes the charge distribution within the cavities, and thus the adsorptive behaviour and the catalytic activity. For these reasons it has become important to determine the cation positions within the framework and much research has been devoted to this in recent years.

The balancing cations in a zeolite can have more than one possible location in the structure. Figure 5.5 shows the available sites in the K^+ form of zeolite-A. Some sites occupy most of the centres of the 6-rings, while the others are in the 8-ring entrances to the β-cages. The presence of cations in these positions effectively reduces the size of the rings and cages to any guest molecules that are trying to enter. In order to alter a

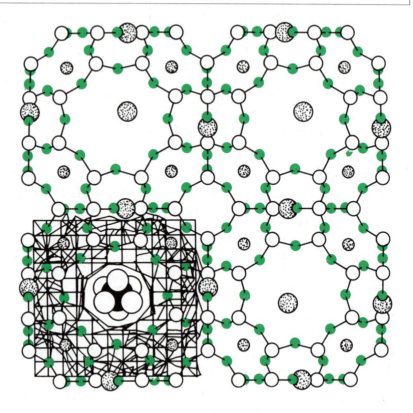

● Oxygen

○ Si/Al

◉ Cation site

Figure 5.5 Framework and cation sites in the K$^+$ form of zeolite-A. In the left-hand corner a molecule of ethane is shown in the channel.

zeolite to allow organic molecules (for instance) to diffuse into or through the zeolite, a divalent cation could be exchanged for a univalent species, thus halving the number of cations present. The divalent cations tend to occupy the sites in the 6-rings, thus leaving the channels free.

Figure 5.6 shows the principal cation sites in the mineral faujasite (and thus by analogy, in zeolite-X and -Y). The principal sites are called S(I), S(I'), S(II), and S(II'). The S(I) sites are in the hexagonal prisms and are usually occupied by ions preferring higher coordination numbers. The S(I') sites are immediately adjacent in the β-cages: S(I) and S(I')

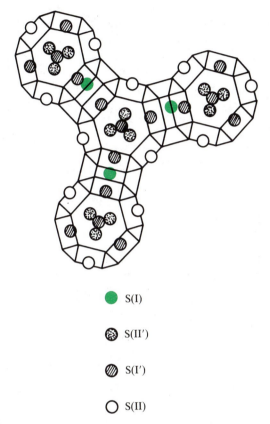

● S(I)

⊛ S(II')

◓ S(I')

○ S(II)

Figure 5.6 Faujasite framework (oxygens not shown) with principal cation sites. S(I') and S(II') are within the β-cages. S(I) and S(II) are in the hexagonal biprisms and the supercages, respectively.

are not simultaneously occupied. The S(II) sites are in the wall of the supercage and are almost all occupied. The S(II') sites are not occupied to any significant extent.

The normal crystalline zeolites contain water molecules which are coordinated to the exchangeable cations. These structures can be dehydrated by heating under vacuum, and in these circumstances the cations move position at the same time, frequently settling on sites with a lower coordination number. The dehydrated zeolites are extremely good drying agents, absorbing water to get back to the preferred hydrated condition.

5.2.4 Cavities and channels

The important structural feature of zeolites, which can be exploited for various uses, is the network of linked **cavities** or **pores** forming a system of **channels** throughout the structure. These cavities are of molecular dimensions and can adsorb species small enough to gain access to them. A controlling factor in whether molecules can be adsorbed in the cavities is the size of the **window** or **pore-opening** into the channel. Hence the importance of the number of tetrahedra forming the window, i.e. the ring size.

Figure 5.4 shows how the window size can vary. A cavity in sodalite is bounded by a 4-ring with a diameter of 260 pm: although this is a very small opening it can admit water molecules. The pore-opening in zeolite-A is 410 pm, determined by an 8-ring window; this is still considerably smaller than the diameter of the internal cavity, which measures 1140 pm in diameter. Faujasite has a 12-ring window of diameter 740 pm, leading into the α-cage of diameter 1180 pm. Values for these and other zeolites are shown in Table 5.1.

Table 5.1 Window and cavity diameters in zeolites

Zeolite	No. of tetrahedra in ring	Window diameter (pm)	Cavity diameter (pm)
Sodalite	4	260	
Zeolite-A	8	410	1140
Erionite-A	8	360 × 520	
ZSM-5	10	510 × 550	
		540 × 560	
Faujasite	12	740	1180
Mordenite	12	670 × 700	
		290 × 570	

The windows to the channels thus form a three-dimensional sieve with mesh widths between about 300 and 1000 pm, hence the well-known name **molecular sieve** for these crystalline aluminosilicates. Zeolites thus have large internal surface areas and high sorption capacities for molecules small enough to pass through the window into the cavities. They can be used to separate mixtures such as straight-chain and branched-chain hydrocarbons.

The zeolites fall into three main categories. The tunnels may be parallel to: (a) a single direction, so that the crystals are **fibrous**; (b) two directions arranged in planes, so that the crystals are **lamellar**; or (c) three directions, such as cubic axes, in which there is strong bonding in three directions. The most symmetrical structures have cubic symmetry. By no means do all zeolites fall neatly into this classification; some, for instance, have a dominant two-dimensional structure interlinked by smaller channels.

A typical fibrous zeolite is **edingtonite**, $Ba[(AlO_2)_2(SiO_2)_3] \cdot 4H_2O$,

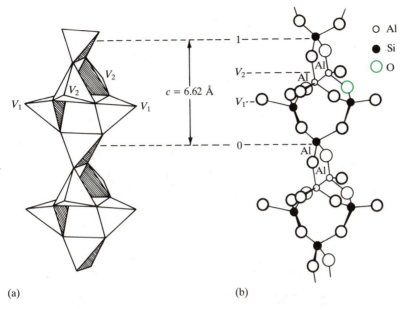

O Al
● Si
◯ O

(a)   (b)

Figure 5.7 The common structural feature of the fibrous zeolites, the chain of linked tetrahedra. It is attached to neighbouring chains by the vertices V_1 and V_2.

which has a characteristic chain formed by the regular repetition of five tetrahedra. The chains link together through oxygen atoms but the concentration of atoms within the chain gives the crystal its fibrous character (Figure 5.7).

Lamellar zeolites occur frequently in sedimentary rocks, e.g. **phillipsite**, $(K/Na)_5[(SiO_2)_{11}(AlO_2)_5] \cdot 10H_2O$ (Figure 5.8), which has channels running parellel to the a-crystallographic axis.

In Figure 5.4b the eight sodalite units illustrated enclose a larger cavity. An alternative way of viewing the crystal structure is to consider the shape of these cavities and how they connect. The cavity has a shape known as a **truncated cuboctahedron**, a cuboctahedron with its corners missing (Figure 5.3c). Figure 5.9 shows the zeolite-A structure again, but with the cavity now outlined in colour. The truncated cuboctahedra are also space-filling (each shares its octagonal face with six others), forming channels which run parallel to the three cubic axial directions through these large cavities.

In the mid-1970s some completely novel zeolite structures were synthesized which led to significant new developments. This family of framework structures comprising the zeolites synthesized by the oil company Mobil, ZSM-5 and ZSM-11, and silicalite and some closely related natural zeolites, have been given the generic name **pentasil**.

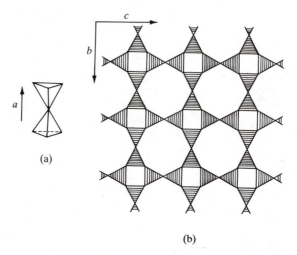

Figure 5.8 The structure of a lamellar zeolite (phillipsite). The double tetrahedral unit shown in (a) projects in the plan (b) as a single (equilateral) triangle.

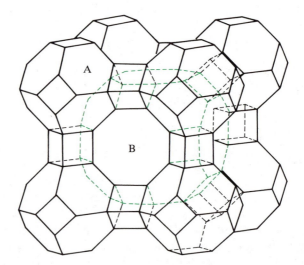

Figure 5.9 Zeolite-A structure showing the truncated cuboctahedron shape of the supercage.

ZSM-5 is a catalyst now widely used in the industrial world. Its structure is generated from the pentasil unit shown in Figure 5.10 (as are the others of this group). These units link into chains, which join to make layers. Appropriate stacking of these layers gives the various pentasil structures.

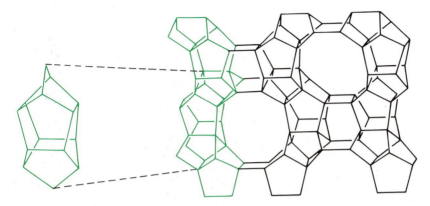

Figure 5.10 A pentasil unit (in colour) together with a slice of the structure of ZSM-5, showing a linked chain of pentasil units highlighted in colour.

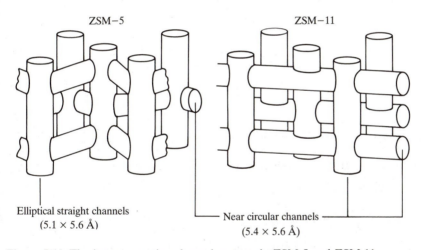

Figure 5.11 The interconnecting channel systems in ZSM-5 and ZSM-11.

Both ZSM-5 and ZSM-11 are characterized by channels controlled by 10-ring windows with diameters of about 550 pm. The pore systems in these zeolites do not link big cavities, but they do contain intersections where larger amounts of free space are available for molecular interactions to take place. Figure 5.11a shows the pore system of ZSM-5 with nearly circular zig-zag channels intersecting with straight channels of elliptical cross-section. This contrasts with the ZSM-11 structure which just has intersecting straight channels of almost circular cross-section (Figure 5.11b).

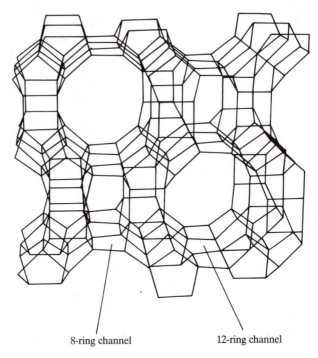

8-ring channel   12-ring channel

Figure 5.12 The channel structure of mordenite.

The channel system for **mordenite** is shown in Figure 5.12. There are two types of channels, governed by 8- and 12-ring windows, respectively, all running parallel to each other and interconnected only by smaller 5- and 6-ring systems.

5.3 PREPARATION OF ZEOLITES

Zeolites are prepared from solutions containing sodium silicates and aluminates, $[Al(OH)_4]^-$, at high pH, obtained by using an alkali metal hydroxide and/or an organic base. A gel forms by a process of co-polymerization of the silicate and aluminate ions. The gel is then heated gently (at 60–100°C) in a closed vessel for about 2 days, producing a condensed zeolite. (These are called hydrothermal conditions.) The product obtained is determined by the synthesis conditions: temperature, time, pH and mechanical movement are all possible variables. The presence of organic bases is useful for synthesizing silicon-rich zeolites.

The formation of novel silicon-rich synthetic zeolites has been facilitated by the use of **templates** such as large quaternary ammonium cations instead of Na^+. For instance, the tetramethylammonium cation,

$[(CH_3)_4N]^+$, is used in the synthesis of ZK-4. The aluminosilicate framework condenses around this large cation, which can subsequently be removed by chemical or thermal decomposition. ZSM-5 is produced in a similar way using tetra-n-propyl ammonium. Only a limited number of large cations can fit into the zeolite framework, and this severely reduces the number of $[AlO_4]$ tetrahedra that can be present, producing a silicon-rich structure.

The preparation of silicon-rich zeolites such as zeolite-Y, can be achieved by varying the composition of the starting materials but can also be done by subsequent removal of aluminium from a synthesized aluminosilicate framework using a chemical treatment. There are several different methods available, one being extraction of the aluminium by mineral acid and another, extraction using complexing agents.

5.4 STRUCTURE DETERMINATION

The structures of the zeolite frameworks have been determined by X-ray and neutron crystallographic techniques. Some of the naturally occurring minerals were characterized in the 1930s and the synthetic zeolites have been investigated from 1956 onwards. Unfortunately, it is extremely difficult for diffraction techniques to determine a structure unequivocally: X-rays for instance are scattered by the electron cloud surrounding a nucleus, and because Al and Si are next to each other in the Periodic Table they scatter electrons almost equally strongly; they are said to have very similar **atomic scattering factors**. The consequence of this is that Al and Si atoms are virtually indistinguishable on the basis of crystallographic data. It is possible to build up a picture of the overall shape of the framework with accurate atomic positions but not to decide which atom is Si and which is Al.

The positions of Si and Al atoms have always been assigned by applying **Loewenstein's rule** which forbids the presence of an Al–O–Al linkage in the structure. A corollary of this rule is that when the Si/Al ratio is 1, the amount of aluminium in a zeolite is maximum and the Si and Al atoms alternate throughout this structure.

When it comes to locating the cations, other problems arise. Not every cation site is completely occupied, so although the cation sites can be located, their occupancy is averaged. Furthermore, zeolites are usually microcrystalline and for successful diffraction studies larger single crystals are needed (although Rietveld powder techniques have been successfully applied, especially with neutrons). One of the techniques currently being used to elucidate zeolite structures successfully is **magic angle spinning NMR spectroscopy (MAS NMR)**. This technique eliminates the broadening of the NMR signals normally observed in solids. The line broadening is

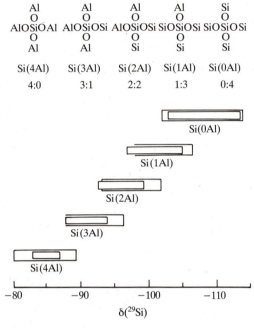

Figure 5.13 The five possible local environments of a silicon atom together with their characteristic chemical shift ranges. The inner boxes represent the ^{29}Si shift ranges suggested in the earlier literature. The outer boxes represent the extended ^{29}Si shift ranges which are more unusual.

due to various anisotropic interactions all of which contain a $(3\cos^2\theta - 1)$ term. When $\cos\theta = (1/3)^{1/2}$, i.e. $\theta = 54°44'$, this term becomes zero. Spinning the sample about an axis inclined at this so-called **magic angle** to the direction of the magnetic field eliminates these sources of broadening, and thus improves the resolution in chemical shift of the spectra.

^{29}Si has a nuclear spin $I = \frac{1}{2}$, and so gives sharp spectral lines with no quadrupole broadening or asymmetry; the sensitivity is quite high and ^{29}Si has a natural abundance of 4.7%.

Pioneering work using MAS NMR on zeolites was carried by E. Lippmaa and G. Engelhardt in the late 1970s. They showed that up to five peaks could be observed for the ^{29}Si spectra of various zeolites and that these corresponded to the five different Si environments that can exist. Each Si is coordinated by four oxygen atoms, but each oxygen can then be attached either to a Si or to an Al atom giving the five possibilities: $Si(OAl)_4$, $Si(OAl)_3(OSi)$, $Si(OAl)_2(OSi)_2$, $Si(OAl)(OSi)_3$, and $Si(OSi)_4$. They also showed, most importantly, that characteristic ranges of these shifts could be assigned to each coordination type. These ranges

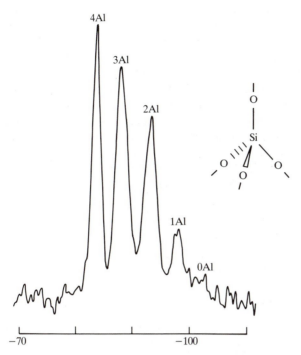

Figure 5.14 ^{29}Si MAS NMR spectrum at 79.6 MHz of the zeolite, analcite, showing five absorptions characteristic of the five possible permutations of Si and Al atoms attached at the corners of the SiO$_4$ tetrahedron as indicated.

could then be used in further structural investigations of other zeolites. These ranges are shown in Figure 5.13. An MAS NMR spectrum of the zeolite analcite is shown in Figure 5.14. Analcite has all five possible environments. Even with this information it is still an extremely complicated procedure to decide where each linkage occurs in the structure. The cation positions can give useful information as they tend to be as close as possible to the negatively charged Al sites.

^{27}Al has a 100% natural abundance and a nuclear spin $I = \frac{5}{2}$, resulting in a strong resonance which is broadened and rendered asymmetric by second-order quadrupolar effects. Because Loewenstein's rule precludes Al–O–Al linkages, every tetrahedral Al environment is Al(OSi)$_4$ and consequently only a single resonance is observed, although the Al chemical shifts of individual zeolites have characteristic values. However, determining the ^{27}Al MAS NMR spectrum of a zeolite can have great diagnostic value as it distinguishes between three different types of coordination: octahedrally coordinated Al, [Al(H$_2$O)$_6$]$^{3+}$, is frequently trapped as a cation in the pores and gives a peak at about 0 ppm (with

$[Al(H_2O)_6]^{3+}$(aq) used as the reference), whereas tetrahedrally co-ordinated Al in the framework gives a peak in the range 50–65 ppm, and tetrahedral $[AlCl_4]^-$ gives a peak at about 100 ppm. Such a peak can occur when a zeolite has been treated with $SiCl_4$ in order to increase the Si/Al ratio in the framework and should disappear with washing.

An important feature of ^{29}Si MAS NMR spectra is that measurement of the intensity of the observed peaks allows the Si/Al ratio of the framework of the sample to be calculated. This can be extremely useful when developing new zeolites for catalysis, as much research has con-centrated on making highly siliceous varieties by replacing the Al in the framework. Conventional chemical analysis only gives an overall Si/Al ratio, which includes trapped octahedral Al species and $[AlCl_4]^-$ which has not been washed away. At high Si/Al ratios, ^{27}Al MAS NMR results are more sensitive and accurate and so are preferred.

5.5 USES OF ZEOLITES

5.5.1 Dehydrating agents

Normal crystalline zeolites contain water molecules, which are coor-dinated to the exchangeable cations. As we noted previously, these structures can be dehydrated by heating under vacuum; in these cir-cumstances the cations move position, frequently settling on sites with a much lower coordination number. The dehydrated zeolites are very good drying agents, adsorbing water to get back to the preferred high coordination condition.

5.5.2 Zeolites as ion exchangers

The cations M^{n+} in a zeolite will exchange with others in a surrounding solution. In this way the Na^+ form of zeolite-A can be used as a water softener: the Na^+ ions exchange with the Ca^{2+} ions from the hard water. The water-softener is reusable as it can be regenerated by running through a very pure saline solution; this is a familiar procedure to anyone who has used a dishwasher. Zeolite-A is now added to detergents as a water softener, replacing the polyphosphates which have given concern about possible ecological damage. It is possible to produce drinking water from seawater by desalinating it through a mixture of Ag and Ba zeolites. This is such an expensive process, however, that it is only useful in emergency.

Some zeolites have a strong affinity for particular cations. **Clinoptilolite** is a naturally occurring zeolite which sequesters caesium. It can be used to recover ^{137}Cs from radioactive waste, exchanging its own Na^+ ions

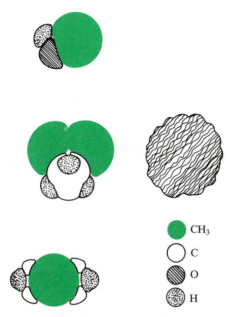

Figure 5.15 Illustration of shape selectivity of ZSM-5. The cross-section of the straight channel can be compared with the size and shape of the molecules shown (approximately viewed along their molecular axis). From top to bottom: methanol, 2,2-dimethyl pentane, *para*-xylene.

for Cs^+ cations. Similarly, zeolite-A can be used to recover radioactive strontium.

5.5.3 Zeolites as adsorbents

Because dehydrated zeolites have very open porous structures, they have large internal surface areas and are capable of adsorbing large amounts of substances other than water. The ring sizes of the windows leading into the cavities determine the size of the molecules that can be adsorbed. An individual zeolite has a highly specific sieving ability that can be exploited for purification or separation. This was first noted for **chabazite** as long ago as 1932, when it was observed it would adsorb and retain small molecules such as formic acid and methanol, but would not adsorb benzene and larger molecules. Chabazite has been used commercially to adsorb polluting SO_2 emissions from chimneys. Similarly, the 410 pm pore-opening in zeolite-A, determined by an 8-ring, (this is much smaller than the diameter of the internal cavity, which is 1140 pm), can admit a methane molecule, but excludes a larger benzene molecule. Computer graphics are extremely useful for illustrating whether a molecule is able

Table 5.2   Applications of molecular sieves in industrial adsorption processes

Fields of application	Uses		
	Drying	Purification	Separations
Refineries and petrochemical industry	Paraffins, olefins, acetylenes, reformer gas, hydrocracking gas, solvents	Sweetening* of 'liquid petrol gas' and aromatics, removal of CO_2 from olefin-containing gases, purification of synthesis gas	Normal and branched-chain alkanes
Industrial gases	H_2, N_2, O_2, Ar, He, CO_2, natural gas	Sweetening and CO_2 removal from natural gas, removal of hydrocarbons from air, preparation of protective gases	Aromatic compounds
Industrial furnaces	Exogas, cracking gas, reformer gas	Removal of CO_2 and NH_3 from exogas and from ammonia fission gas	Nitrogen and oxygen

* 'Sweetening' is the removal of sulphur-containing compounds.

to pass through a pore-opening or channel (Figure 5.15). The atoms are assigned their van der Waals' radii for this purpose.

The zeolites that are useful as molecular sieves do not show an appreciable change in the basic framework structure on dehydration, although the cations move to positions of lower coordination. After dehydration, zeolite-A and others are remarkably stable to heating and do not decompose below about 700°C. The cavities in dehydrated zeolite-A amount to about 50% of the volume.

There are many uses of the selective sorbent properties of zeolites of which we will describe just a few. Table 5.2 gives a brief summary of the industrial uses that are made. The zeolites are regenerated after use by heating, evacuation or flushing with pure gases.

It is possible to 'fine-tune' the pore-opening of a zeolite to allow the adsorption of specific molecules. As discussed earlier, one method is to change the cation. This method can be used to modify zeolite-A so as to separate branched and cyclic hydrocarbons from the straight-chain

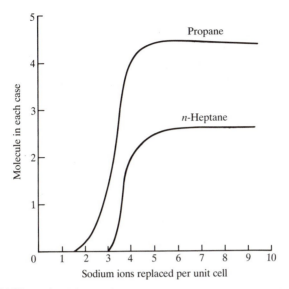

Figure 5.16 Effect of calcium exchange for sodium in zeolite-A on hydrocarbon adsorption. Replacement of four sodium ions by two calcium ions permits easy diffusion of *n*-alkanes into the zeolite channels.

alkanes (paraffins). When Na^+ ions are replaced by Ca^{2+} ions, the effective aperture increases. Once approximately one-third of the Na^+ ions have been replaced by Ca^{2+}, many straight chain alkanes can be adsorbed (Figure 5.16). However, all branched-chain, cyclic and aromatic hydrocarbons are excluded because their cross-sectional diameters are too large. (Figure 5.5 shows an ethane molecule adsorbed in one of the supercages.) This process can be useful industrially for separating the long straight-chain hydrocarbons required as the starting materials in the manufacture of biodegradable detergents. Petrol can also be upgraded by removing from it the straight-chain, low-octane-number alkane constituents which produce 'pinking' ('knocking'), small explosions which are harmful to engines.

At −196°C, oxygen is freely adsorbed by zeolite-A (Ca form) whereas nitrogen is essentially excluded. The two molecules are not very different in size: O_2 has a diameter of 346 pm whereas that of N_2 is 364 pm. As the temperature rises the adsorption of N_2 increases to a maximum at around −100°C. The main reason is thought to be due to the thermal vibrations of the oxygen atoms in the window. Over a range of 80–300 K a variation of vibrational amplitude of 10–20 pm could well be expected and thus a variation of 30 pm in the window diameter would not be unreasonable. This would make the window just small enough at lower temperatures to exclude the N_2 molecules. The other method used for 'fine-tuning' the pore-openings is to change the Si/Al ratio. An increase in the proportion of Si will: (a) slightly decrease the unit cell size and thus the size of the cavities; (b) decrease the number of cations, thus freeing the channels; and (c) make the zeolite more hydrophobic (literally 'water-hating') in character. Hydrophobic zeolites can potentially be used to remove organic molecules from aqueous solution; possible uses range from the removal of toxic materials from blood, to the production of non-alcoholic beverages by the selective removal of alcohol, and to the decaffeination of coffee.

5.5.4 Zeolites as catalysts

Zeolites are very useful catalysts displaying several important properties that are not found in traditional amorphous catalysts. Amorphous catalysts have always been prepared in a highly divided state in order to give a high surface area and thus a large number of catalytic sites. The presence of the cavities in zeolites provides a very large internal surface area that can accommodate as many as 100 times more molecules than the equivalent amount of amorphous catalyst. Zeolites are also crystalline and so can be prepared with improved reproducibility: they tend not to show the varying catalytic activity of amorphous catalysts. Furthermore their molecular sieve action can be exploited to control which molecules have access to (or which molecules can depart from) the active sites. This is generally known as **shape-selective catalysis**.

The catalytic activity of decationized zeolites is attributed to the presence of acidic sites arising from the $[AlO_4]$ tetrahedral units in the framework. These acid sites may be Brønsted or Lewis in character. Zeolites as normally synthesized usually have Na^+ ions balancing the framework charges, but these can be readily exchanged for protons by direct reaction with an acid, giving surface hydroxyl groups – the **Brønsted sites**. Alternatively, if the zeolite is not stable in acid solution, it is common to form the ammonium, NH_4^+, salt, and then heat it so that ammonia is driven off, leaving a proton. Further heating removes water from the Brønsted site, exposing a tricoordinated Al ion, which has electron-pair acceptor properties; this is identified as a **Lewis acid site**. A scheme for the formation of these sites is shown in Figure 5.17. The surfaces of zeolites can thus display either Brønsted or Lewis acid sites, or both depending on how the zeolite is prepared. Brønsted sites are converted into Lewis sites as the temperature is increased above 600°C.

Not all zeolite catalysts are used in the decationized or acid form; it is also quite common to replace the Na^+ ions with lanthanide ions such as La^{3+} or Ce^{3+}. These ions now place themselves so that they can best neutralize three separated negative charges on tetrahedral Al in the framework. The separation of charges causes high electrostatic field gradients in the cavities which are sufficiently large to polarize C—H

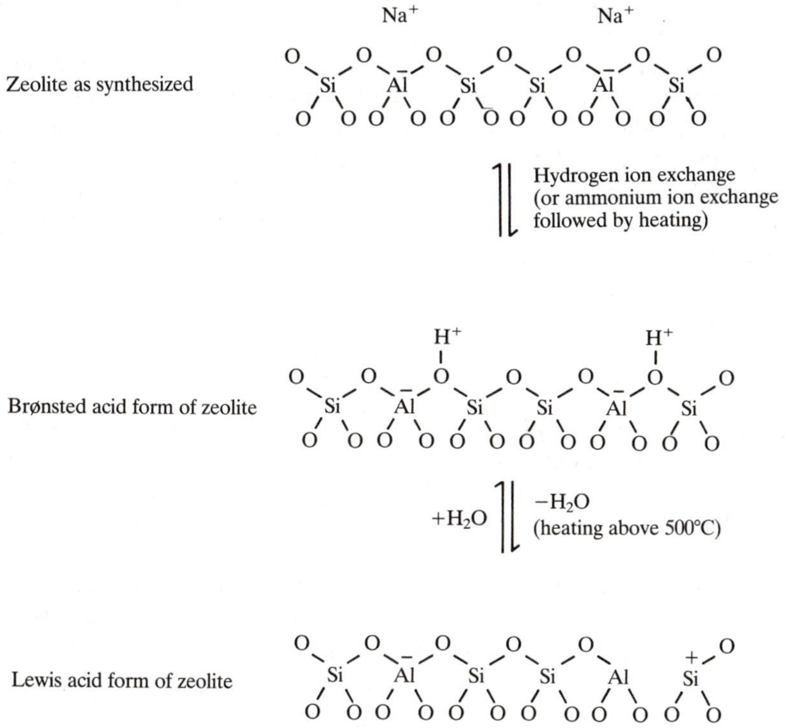

Figure 5.17 Scheme for the generation of Brønsted and Lewis acid sites in zeolites.

bonds or even to ionize them, enabling reaction to take place. This effect can be strengthened by a reduction in the Al content of the zeolite so that the [AlO$_4$] tetrahedra are farther apart. If one thinks of a zeolite as a solid ionizing solvent, the difference in catalytic performance of various zeolites can be likened to the behaviour of different solvents in solution chemistry. It was a rare-earth substituted form of zeolite-X that became the first commercial zeolite catalyst for the cracking of petroleum in the 1960s. Crude petroleum is initially separated by distillation into fractions, and the heavier gas-oil fraction is cracked over a catalyst to give petrol (gasoline). A form of zeolite-Y, that has proved more stable at high temperatures, is now used. These catalysts yield ~20% more petrol than earlier catalysts, and do so at lower temperatures. The annual catalyst usage for catalytic cracking is worth billions of £s.

A third way of using zeolites as catalysts is to replace the Na$^+$ ions with other metal ions such as Ni^{2+}, Pd^{2+}, or Pt^{2+} and then reduce them *in situ* so that metal atoms are deposited within the framework. The resultant material displays the properties associated with a supported metal catalyst and extremely high dispersions of the metal can be achieved. Another technique for preparing a zeolite-supported catalyst involves the physical adsorption of a volatile inorganic compound followed by thermal decomposition: $Ni(CO)_4$ can be adsorbed on zeolite-X and decomposed with gentle heating to leave a dispersed phase of nearly atomic nickel in the cavities, and this has been found to be a good catalyst for the conversion of carbon monoxide to methane:

$$CO + 3H_2 = CH_4 + H_2O.$$

There are three types of shape-selective catalysis.

1. **Reactant-selective catalysis**: only molecules with dimensions less than a critical size can enter the pores and reach the catalytic sites, and so react there. This is illustrated diagrammatically in Figure 5.18a in which a straight-chain hydrocarbon is able to enter the pore and react but the branched-chain hydrocarbon is not.
2. **Product-selective catalysis**: only products less than a certain dimension can leave the active sites and diffuse out through the channels, as illustrated in Figure 5.18b for the preparation of xylene. A mixture of all three isomers is formed in the cavities but only the *para* form is able to escape.
3. **Transition-state-selective catalysis**: certain reactions are prevented because the transition state requires more space than is available in the cavities, as shown in Figure 5.18c for the transalkylation of dialkylbenzenes.

Reactant-shape-selective catalysis is demonstrated in the dehydration of butanols. If butan-1-ol (*n*-butanol) and butan-2-ol (*iso*-butanol) are dehydrated either over zeolite-A or over zeolite-X (both in the Ca form) we see a difference in the products formed.

$$CH_3CH_2CH_2CH_2OH = CH_3CH_2CH{=}CH_2 + H_2O$$

$$CH_3CH_2\underset{\underset{\displaystyle OH}{|}}{C}HCH_3 = CH_3CH{=}CHCH_3 + H_2O$$

Zeolite-X has windows large enough to admit both alcohols easily, and both undergo conversion to the corresponding alkene. Over zeolite-A, however, the dehydration of the straight-chain alcohol is straightforward but virtually none of the branched-chain alcohol is converted, as it is too large to pass through the smaller windows of zeolite-A. These results are summarized in Figure 5.19. Notice that, at higher temperatures, curve d begins to rise. This is because the lattice vibrations increase with

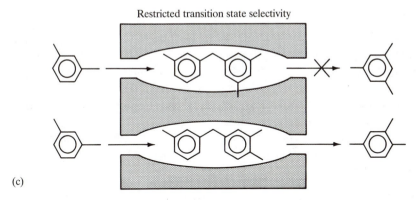

Reactant selectivity

(a)

Product selectivity

CH₃OH +

(b)

Restricted transition state selectivity

(c)

Figure 5.18 Shape-selective catalysis: (a) reactant, (b) product and (c) transition state.

temperature, making the pore-opening slightly larger and thus beginning to admit butan-2-ol. The very slight conversion at lower temperatures is thought to take place on external sites.

In the acid-catalysed transalkylation of dialkylbenzenes, one of the alkyl groups is transferred from one molecule to another. This bimolecular reaction involves a diphenylbenzene transition state. When the transition state collapses it can split to give either the 1,2,4-isomer or the

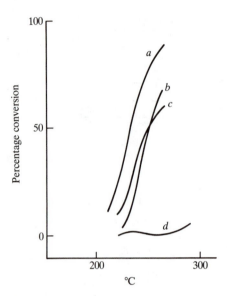

Figure 5.19 Dehydration of (a) *iso*-butanol on Ca-X, (b) *n*-butanol on Ca-X, (c) *n*-butanol on Ca-A, and (d) *iso*-butanol on Ca-A.

1,3,5-isomer, together with the monoalkylbenzene. When the catalyst used for this reaction is mordenite (Figure 5.18c), the transition state for the formation of the symmetrical 1,3,5-isomer is too large for the pores, and the 1,2,4-isomer is formed in almost 100% yield. (This compares with the equilibrium mixtures, in which the symmetrically substituted isomers tend to dominate.)

One of the industrial processes using ZSM-5 provides us with an example of product-shape-selective catalysis: the production of 1,4-(*para*-)xylene. *Para*-xylene is used in the manufacture of terephthalic acid, the starting material for the production of polyester fibres such as 'Terylene'.

Xylenes are produced in the alkylation of toluene by methanol:

$$CH_3OH \;+\; \overset{CH_3}{\bigcirc} \;=\; \overset{CH_3}{\bigcirc}\!\!^{CH_3} \;+\; \overset{CH_3}{\underset{CH_3}{\bigcirc}} \;+\; \overset{CH_3}{\underset{CH_3}{\bigcirc}} \;+\; H_2O$$

methanol   toluene   (*ortho*-)   (*meta*-)   (*para*-)
                       1,2-xylene   1,3-xylene   1,4-xylene

p-Xylene in ZSM−5

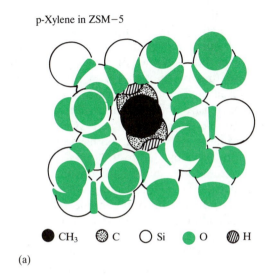

● CH₃   ◉ C   ○ Si   ● O   ◍ H

(a)

m-Xylene in ZSM−5

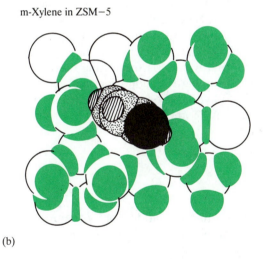

(b)

Figure 5.20 Computer graphics representations, showing how (a) *para*-xylene fits neatly in the pores of ZSM-5 whereas (b) *meta*-xylene cannot diffuse through its channels.

The selectivity of the reaction over ZSM-5 occurs because of the difference in the rates of diffusion of the different isomers through the channels. This is confirmed by the observation that selectivity increases with increasing temperature, indicating the increasing importance of diffusion limitation. The diffusion rate of *para*-xylene is approximately 1000 times faster than that of the other two isomers. The computer graphics representation in Figure 5.20 shows why *meta*- and *ortho*-xylene do not fit in the pores easily and so cannot diffuse along them. The xylenes isomerize within the pores, and so *para*-xylene diffuses out while the *ortho*- and *meta*- isomers are trapped and have more time to convert to the *para*- form before escaping. Up to 97% of selective conversion to *para*-xylene has been achieved by suitable treatment of the catalyst.

ZSM-5 is used as a catalyst in the disproportionation of toluene (a by-product of petroleum refining) to form benzene and *para*-xylene, both of which are more valuable products. ZSM-5 is also the catalyst used to convert methanol into hydrocarbons. This research received a great impetus at the end of the 1970s when oil was in short supply and prices rose sharply. Subsequently, more oil was released and prices dropped; consequently much of this research was put into abeyance, and plants were left at the pilot stage. It may prove to be an important process for those countries, such as New Zealand, with no oil reserves of their own, but which do have a supply of natural gas (CH_4). The methane is first converted to methanol, CH_3OH, which is then partially broken down into dimethyl ether and water. (Alternatively the methanol feed can be readily synthesized from carbon monoxide and hydrogen.) The conversion over ZSM-5 into hydrocarbons produces mainly branched chain and aromatic hydrocarbons in the C_9–C_{10} range, which is ideal for high octane unleaded fuel.

5.6 POSTSCRIPT

We end this chapter with a quotation from a lecture given by Prof. J.M. Thomas. It should please those of you who are beginning to love the patterns and symmetry of crystalline solids.

> ... Figure 5.21 shows, on the right, a projected structure of the zeolite we have been studying, a synthetic catalyst discovered in New Jersey in 1975. Its structure was elucidated some 6 years ago. On the left, I reproduce a pattern made on the wall of a mosque in Baku in the Soviet state of Azerbaijan in 1086 AD, the year that the Domesday Book appeared by order of William the Conqueror. These two structures have exactly the same pattern.
>
> 'There is nothing new under the sun.'

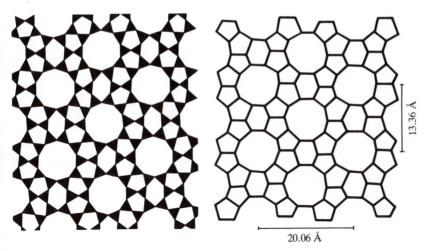

13.36 Å

20.06 Å

Figure 5.21 Around a central 10-membered ring in the synthetic zeolite catalyst shown schematically on the right, there are eight five-membered and two six-membered rings. In exactly the same sequence, such five- and six-membered rings circumscribe a central 10-membered ring in the pattern on the left.

FURTHER READING

Few solid state books cover this subject. The following are specialist books and articles in the field that contain accessible material.

Breck, D.W. (1974) *Zeolite Molecular Sieves*, John Wiley, New York.

Barrer, R.M. (1982) *Hydrothermal Chemistry of Zeolites*, Academic Press, New York.

Dyer, A. (1988) *An Introduction to Zeolite Molecular Sieves*, John Wiley, New York.

Whan, D.A. (1981) Structure and catalytic activity of zeolites, *Chemistry in Britain*, 532–5.

Fyfe, C.A., Thomas, J.M., Klinowski, J. and Gobbi, G.C. (1983) Magic-angle-spinning NMR spectroscopy and the structure of zeolites, *Ang Chem (Int Ed)*, **22**, 259–336.

Ramdas, S., Thomas, J.M., Betteridge, P.W., Cheetham, A.K. and Davies, E.K. *Modelling the chemistry of zeolites by computer graphics* (1984) *Ang Chem (Int Ed)*, **23**, 671–9.

Csicsery, S.M. (1985) Shape selective catalysis in zeolites, *Chemistry in Britain*, 473–7.

Kerr, G.T. (1989) Synthetic zeolites, *Scientific American*, July, 82–7.

QUESTIONS

1. Figure 5.22 shows the ^{29}Si MAS NMR spectrum of faujasite. Use this figure to determine which Si environments are most likely to be present.

2. Figure 5.23 shows the ^{29}Si spectrum of the same sample of faujasite as Figure 5.22, but after treatment with $SiCl_4$ and washing with water. What has happened to it?

3. A sample of faujasite was treated with $SiCl_4$ and four ^{27}Al MAS NMR spectra were taken at various stages afterwards (Figure 5.24). Describe carefully what has happened during the process.

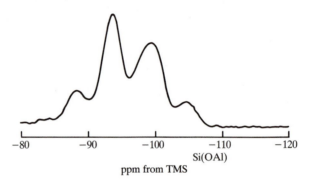

Figure 5.22 ^{29}Si MAS NMR spectrum at 79.6 MHz of faujasite of Si/Al = 2.61 (zeolite-Y).

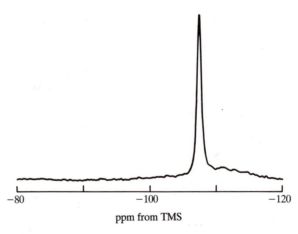

Figure 5.23 ^{29}Si MAS NMR spectrum at 79.6 MHz of faujasite of Si/Al = 2.61 after successive dealuminations with $SiCl_4$ and washings.

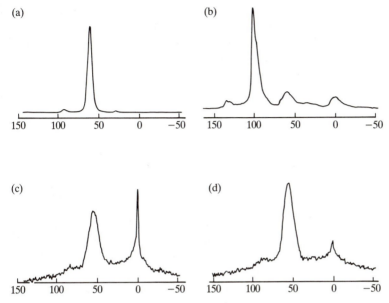

Figure 5.24 ^{27}Al MAS NMR spectra at 104.2 MHz obtained on zeolite-Y samples at various stages of the SiCl$_4$ dealumination procedure. (a) Starting faujasite sample. (b) Intact sample after reaction with SiCl$_4$ before washing. (c) Sample (b) after washing with distilled water. (d) After several washings.

4. Zeolite-A has a single peak in the ^{29}Si MAS NMR spectrum at 89 ppm and has a Si/Al ratio of 1. Comment on these observations.

5. Zeolite-A (Ca form), when loaded with platinum, has been found to be a good catalyst for the oxidation of hydrocarbon mixtures. However, if the mixture contains branched-chain hydrocarbons, these do not react. Suggest a possible reason.

6. Both ethene and propene can diffuse into the channels of a particular mordenite catalyst used for hydrogenation. Explain why only ethane is produced.

7. Explain why, when 3-methyl pentane and *n*-hexane are cracked over zeolite-A (Ca form) to produce smaller hydrocarbons, the percentage conversion for 3-methyl pentane is less than 1% whereas that for *n*-hexane is 9.2%.

8. When toluene is alkylated by methanol with a ZSM-5 catalyst, increase in the crystallite size from 0.5 to 3 µm approximately doubles the amount of *para*-xylene produced. Suggest a possible explanation.

9. The infrared stretching frequency of the hydroxyl associated with the Brønsted sites in decationized zeolites, falls in the range 3600–3660 cm^{-1}. As the Si/Al ratio in the framework increases, this frequency tends to decrease. What does this suggest about the acidity of the highly siliceous zeolites?

ANSWERS

1. The peaks in the spectrum maximize at approximately -88, -93, -99 and -105 ppm. If you mark these values on the chart in Figure 5.13, then you will see that the best correspondence is to the four linkages: Si(OAl)$_3$(OSi), Si(OAl)$_2$(OSi)$_2$, Si(OAl)(OSi)$_3$, and Si(OSi)$_4$.

2. There is only one peak in the new spectrum at approximately -108 ppm. Reference to the chart in Figure 5.13 suggests that this is due to a Si(OSi)$_4$ environment. Clearly the treatment with SiCl$_4$ has removed all the tetrahedral Al from the framework. The intensity measurements on this peak indicate a Si/Al ratio of 55.

3. The starting sample (a) clearly shows the presence of tetrahedral Al in the framework (peak at 61 ppm). After treatment with SiCl$_4$ (b) the amount of Al in the framework has been reduced considerably but there is a very strong peak due to [AlCl$_4$]$^-$ (at 100 ppm) and also a peak due to octahedral aluminium at 0 ppm. The first washings (c) remove Na$^+$[AlCl$_4$]$^-$ from the sample and repeated washing (d) also removes some of the octahedrally coordinated Al.

4. The chart in Figure 5.13 shows that the most likely Si environment for zeolite-A is Si(OAl)$_3$(OSi). However, we know that the Si/Al ratio is 1, so this coordination is not possible without systematically breaking Loewenstein's rule. It was spectroscopic work on this structure (and on zeolite ZK-4) that eventually confirmed the Si(OAl)$_4$ structure of zeolite-A with strict alternation of Si and Al and led to the extended ranges shown in the chart.

5. This system demonstrates reactant-shape-selective catalysis. The branched hydrocarbons are too bulky to pass through the pore-openings in the catalyst.

6. We see here product-shape-selective catalysis in operation: both reactant molecules are small enough to diffuse into the pores and be hydrogenated, but the slightly larger propane molecule cannot leave.

7. Zeolite-A (Ca) shows reactant selectivity. The straight-chain n-hexane can pass through the windows and undergo reaction but the branched-chain 3-methylpentane is excluded. The selective cracking of straight-

chain hydro-carbons in the presence of branched chains is an important industrial process known as **selectoforming**, which improves the octane number of the fuel.

8. There are two possible reasons for the increased percentage of *para*-xylene. First, the increased crystal size increases the distance travelled by the diffusing molecules along the channels, giving more time for isomerization to the *para*-form. Second, the shape-selective reactions take place in the internal pores of the zeolites, rather than on the surface. Larger crystals have a higher ratio of internal to external sites, and so we would expect selectivity to improve.

9. As the Si/Al ratio increases, the —OH stretching frequency falls. This indicates a decrease in the covalent bond strength making ionization of H^+ easier, i.e. an increase in acid strength.

6 Optical properties of solids

6.1 INTRODUCTION

Perhaps the most well-known example of solid state optical devices is the **laser** and, with the advent of video- and compact discs and laser printers, lasers are becoming almost commonplace. Of interest to the solid state chemist are two types of laser, typified by the **ruby laser** and the **gallium arsenide laser**. Because laser light is more easily modulated than light from other sources, it is increasingly used for sending information; light travelling along optical fibres replaces electrons travelling along wires in, for example, telecommunications. To transmit light over long distances, the optical fibres must have particular absorption and refraction properties and the development of suitable substances has become an important area of research.

Applications of solid state optical properties were known and commercially important before lasers were developed and we also consider examples of these. **Light-emitting diodes (LEDs)** are used for displays on digital watches and scientific instruments. The mechanism by which light is produced in LEDs is similar to that in the gallium arsenide (GaAs) laser. Another important group of solids are the light-emitting solids known as **phosphors** which are used on television screens and fluorescent light tubes.

Broadly speaking, there are two situations that have to be considered in explaining devices such as those we have mentioned. In the first, which is relevant to the ruby laser and to phosphors for fluorescent lights, the light is emitted by an impurity ion in a host lattice. We are concerned here with what is essentially an atomic spectrum modified by the lattice. In the second case, which applies to LEDs and the gallium arsenide laser, the optical properties of the delocalized electrons in the bulk solid are important. We shall start with the first case and take atomic spectra as our starting point.

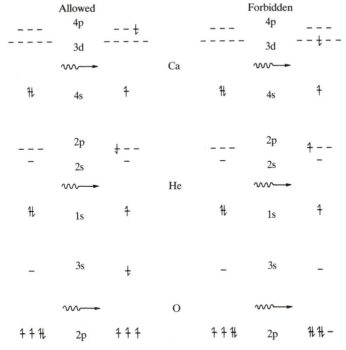

Figure 6.1 Allowed and forbidden atomic transitions.

6.2 THE INTERACTION OF LIGHT WITH ATOMS

When an atom absorbs a photon of light of the correct wavelength, it undergoes a transition to a higher energy level. To a first approximation in many cases, we can think of one electron in the atom absorbing the photon and being excited. The electron will only absorb the photon if the photon's energy matches that of the energy difference between the initial and final electronic energy level, and if certain rules, known as **selection rules**, are obeyed. In light atoms, the electron cannot change its spin and its orbital angular momentum must change by one unit; in terms of quantum numbers $\Delta s = 0$, $\Delta l = \pm 1$. (One way of thinking about this is that the photon has zero spin and one unit of angular momentum. Conservation of spin and angular momentum then produces these rules.) For a sodium atom, for example, the $3s$ electron can absorb one photon and go to the $3p$ level (there is no restriction on changes of the principal quantum number). The $3s$ electron will not, however, go to the $3d$ or $4s$ level. Figure 6.1 illustrates allowed and forbidden transitions.

However, the spin and orbital angular momenta are not entirely

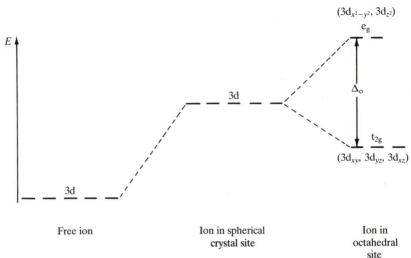

Figure 6.2 Splitting of d levels in an octahedral site in a crystal.

independent and coupling between them allows forbidden transitions to occur; however, the probability of an electron absorbing a photon and being excited to a forbidden level is much smaller than the probability of it being excited to an allowed level. Consequently spectral lines corresponding to forbidden transitions are less intense than those corresponding to allowed transitions.

An electron that has been excited to a higher energy level will sooner or later return to the ground state. It can do this in several ways. The electron may simply emit a photon of the correct wavelength at random some time after it has been excited. This is known as **spontaneous emission**. Alternatively a second photon may come along and instead of being absorbed may induce the electron to emit. This is known as **induced** or **stimulated emission** and plays an important role in the action of lasers. The emitted photon in this case is in phase with and travelling in the same direction as the photon inducing the emission; the resulting beam of light is said to be coherent. Finally the atom may collide with another atom, losing energy in the process, or give energy to its surroundings in the form of vibrational energy. These are examples of non-radiative transitions. Spontaneous and stimulated emission obey the same selection rules as absorption. Non-radiative transitions have different rules. In a crystal (or of course a molecule), the atomic energy levels and selection rules are modified. Let us take as an example an ion with one d electron outside a closed shell (Ti^{3+} for example). This will help us understand the ruby laser.

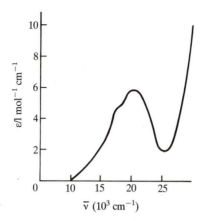

Figure 6.3 The t_{2g} to e_g transition of Ti^{3+}. The band is in fact two overlapping bands. This is due to a further splitting of the e_g levels.

In the free ion, the five $3d$ orbitals all have the same energy. In a crystal these levels are split; for example, if the ion occupied an octahedral hole, the $3d$ levels would be split into a lower, triply degenerate (t_{2g}) level and a higher, doubly degenerate (e_g) level (Figure 6.2). An electronic transition between these levels is now possible. In the free ion a transition from one d level to another involves zero energy change, and so would not be observed even if it were allowed. In the crystal, the transition involves a change in energy, but is still forbidden by the selection rules. Lines corresponding to such transitions can, however, be observed, albeit with low intensity, because the crystal vibrations mix different electronic energy levels. Thus the $3d$ levels may be mixed with the $4p$ giving a small fraction of 'allowedness' to the transition. Figure 6.3 shows an absorption band due to a transition from t_{2g} to e_g for the ion Ti^{3+} which has one d electron. (This band is in fact two closely spaced bands as the excited state is distorted from a true octahedron and the e_g level further split into two.) The role played by a similar forbidden transition in the operation of the ruby laser is described below.

6.2.1 The ruby laser

Ruby is corundum (one form of Al_2O_3) with 0.04–0.5% Cr^{3+} ions as an impurity replacing aluminium ions. The aluminium ions, and hence the chromium ions, occupy distorted octahedral sites. As discussed above, therefore the $3d$ levels of the chromium ion will be split. Cr^{3+} has three $3d$ electrons, and, in the ground state, these occupy separate orbitals with parallel spins. When light is absorbed, one of these electrons can

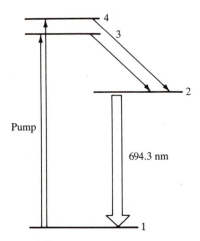

Figure 6.4 The states of the Cr^{3+} ion involved in the ruby laser transition.

undergo a transition to a higher energy $3d$ level; this is similar to the transition discussed in the previous section, but with three eletrons it is necessary to consider changes in the electron repulsion as well as changes in orbital energies. Including electron repulsion changes gives two transitions corresponding to the jump from the lower to the higher $3d$ level.

Having absorbed light and undergone one of these transitions, the chromium ion could now simply emit radiation of the same wavelength and return to the ground state. However, in ruby, there is a fast, radiationless transition in which the excited electron loses some of its energy and the crystal gains vibrational energy. The chromium ion is left in a state in which it can only return to the ground state by a transition in which an electron changes its spin. Such a transition is doubly forbidden since it still breaks the rule that forbids $3d \rightarrow 3d$ and is even less likely to occur than the original absorption process. The states involved are shown schematically in Figure 6.4. The ions absorb light and go to states 3 and 4. They then undergo a radiationless transition to state 2. Because the probability of spontaneous emission for state 2 is low, and there is no convenient non-radiative route to the ground state, a considerable population of state 2 can build up. When eventually (about 5 ms later) some ions in state 2 return to the ground state, the first few spontaneously emitted photons interact with other ions in state 2 and induce these to emit. The resulting photons will be in phase and travelling in the same direction as the spontaneously emitted photons and will induce further emission as they travel through the ruby. The ruby is in a reflecting cavity so that the photons are reflected back into the crystal when they reach the edge. The reflected photons induce further emission and

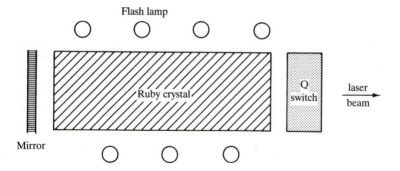

Figure 6.5 Sketch of a ruby laser.

Table 6.1

Ion	Host	Wavelength emitted (nm)
Nd^{3+}	Fluorite (CaF_2)	1046
Sm^{3+}	Fluorite	708.5
Ho^{3+}	Fluorite	2090
Nd^{3+}	Calcium tungstate ($CaWO_4$)	1060

by this means, an appreciable beam of coherent light is built up. The mirror on one end can then be removed and a pulse of light emitted. The name **laser** is a reflection of this build up of intensity. It is an acronym standing for Light Amplification by Stimulated Emission of Radiation. (Similar devices producing coherent beams of microwave radiation are known as masers.) A typical arrangement for a pulsed ruby laser is shown in Figure 6.5. A high intensity flash lamp excites the Cr^{3+} ions from level 1 to levels 3 and 4. The lamp can lie alongside the crystalline rod of ruby or be wrapped around it. At one end of the reflective cavity surrounding the ruby crystal is a Q switch, which switches from being reflective to transmitting the laser light and can be as simple as a rotating segmented mirror, but is usually a more complex device.

Ruby was the first material for lasers but there are now several other crystals that are employed. The crystals used need to contain an impurity with an energy level such that return to the ground state is only possible by a forbidden transition in the infrared, visible or near ultraviolet. It must also be possible to populate this level via an allowed (or at least less forbidden) transition. Research has tended to concentrate on transition metal ions and lanthanide ions in various hosts since these ions have suitable transitions of the right wavelength. Some examples are given in Table 6.1 with the wavelength of the laser emission.

6.2.2 Phosphors in fluorescent lights

Phosphors are solids which absorb energy and re-emit it as light. As in the lasers we have just described the emitter is usually an impurity ion in a host lattice. However, for the uses to which phosphors are put it is not necessary to produce intense, coherent beams of light, and the emitting process is spontaneous rather than induced. There are many applications of phosphors, for example television pictures are produced by phosphors that are bombarded with an electron beam (cathode rays). In terms of tonnage produced, one of the most important applications is fluorescent light tubes.

Fluorescent lights produce radiation in the ultraviolet (254 nm) by passing an electric discharge through a low pressure of mercury vapour. The tube is coated inside with a white powder which absorbs the ultraviolet light and emits visible radiation. For a good fluorescent light, the efficiency of the conversion should be high and the emitted light should be such that the appearance of everyday objects viewed by it should resemble as closely as possible their appearance in daylight. Most phosphors for fluorescent lights have been based on alkaline earth halophosphates such as $3Ca_3(PO_4)_2 \cdot CaF_2$. As in lasers, the usual dopants are transition metal or lanthanide ions, but more than one impurity ion is needed to approximate the whole visible spectrum. Not all the impurity ions need to be capable of absorbing the exciting radiation, however, as the host lattice can act to transmit the energy from one site to another. For example, in a phosphor doped with Mn^{2+} and Sb^{3+} ions, the ultraviolet radiation from the mercury lamp is only absorbed by the antimony (Sb^{3+}) ions. The excited antimony ion drops down to a lower excited state via a non-radiative transition. Emission from this lower state produces a broad band in the blue region of the visible spectrum. Some of the energy absorbed by the antimony travels through the host crystal and is absorbed by the manganese ions. The excited Mn^{2+} ions emit yellow light and return to the ground state. The two emission bands together produce something close to daylight. Phosphors have been introduced which are more efficient and give a closer approximation to daylight. A good approximation is, for example, given by a combination of blue from barium magnesium aluminate doped with divalent europium (Eu^{2+}), green from an aluminate doped with cerium (Ce^{3+}) and terbium (Tb^{3+}) ions, and red from yttrium oxide doped with trivalent europium (Eu^{3+}).

In these phosphors and in the ruby laser, light was absorbed and emitted by electrons localized on an impurity site, but in other optical devices, the radiation is emitted by delocalized electrons. In the next section, therefore, we shall consider the absorption and emission of radiation in solids with delocalized electrons, in particular in semiconductors.

6.3 ABSORPTION AND EMISSION OF RADIATION IN SEMICONDUCTORS

Radiation falling on a semiconductor will be absorbed by electrons in delocalized bands, in particular those near the top of the valence band, causing these electrons to be promoted to the conduction band. Because there are many closely packed levels in an energy band, the absorption spectrum is not a series of lines as in atomic spectra, but a broad peak with a sharp threshold close to the band gap energy. The absorption spectrum of GaAs, for example, is shown in Figure 6.6.

Transitions to some levels in the conduction band are more likely than transitions to others. This is because transitions between valence band and conduction band levels, like those between atomic energy levels, are governed by selection rules. The spin selection rule still holds; when promoted to the conduction band the electron does not change its spin. Orbital angular momentum rules are not appropriate for energy bands, however, and so the rule governing change in the quantum number l is replaced by a restriction on the wave vector, $\mathbf{k}$. As seen in Chapter 2, the energy levels in a band are characterized by the wave vector, the momentum of the electron wave being given by $\mathbf{k}\hbar$. The momentum of a photon with wavelength in the infrared, visible or ultraviolet is very small compared to that of the electron in the band and so conservation of momentum produces the selection rule for transitions between bands. An electron cannot change its wave vector when it absorbs or emits radiation. Thus an electron in the valence band with wave vector, $\mathbf{k_i}$, can only undergo allowed transitions to levels in the conduction band which also have wave vector, $\mathbf{k_i}$. In some solids, for example GaAs, the level at the top of the valence band and that at the bottom of the conduction

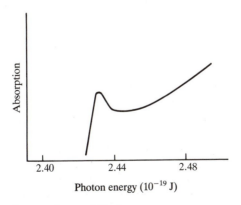

Figure 6.6 Absorption spectrum of GaAs.

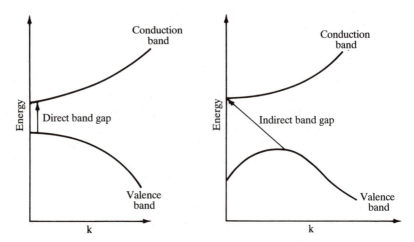

Figure 6.7 Sketch of energy bands for (a) a solid with a direct band gap and (b) a solid with an indirect band gap. Note that in this diagram the horizontal axis is k, *not* the density of states. In this representation a band is shown as a line going from 0 to the maximum value of k occurring for that band.

band have the same wave vector. There is then an allowed transition at the band gap energy. Such solids are said to have a **direct band gap**.

For other semiconductors, for example silicon, the direct transition from the top of the valence band to the bottom of the conduction band is forbidden. These solids are said to have an **indirect band gap**. Illustrations of band structures for solids with direct and indirect band gaps are given in Figure 6.7. Note that in these diagrams, the energy for a band in one direction is plotted against the wave number, k. Similar diagrams were given in Chapter 2 for the free electron model but more accurate calculations show that curves of this type adopt many shapes and do not necessarily resemble those predicted by the free electron model.

The simple free electron model might suggest that the lowest energy orbital in any band is that with $k = 0$. Figure 6.8, however, illustrates two combinations of orbitals that will have $k = 0$ for a chain because all the atomic orbitals are combined in phase. The combination of p orbitals is obviously antibonding and so would be expected to have the highest energy in its band; the combination of s orbitals is bonding and would have the lowest energy in its band. If the p band lay below the s band, a transition between these levels would be allowed and would correspond to a direct transition across the band gap. In real solids, the highest and lowest levels in bands will contain contributions from different types of atomic orbital and it becomes difficult to predict whether a band gap will be direct or indirect. One consequence of an indirect band gap is that an

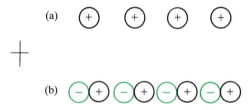

Figure 6.8 A row of (a) s orbitals and (b) p orbitals both with $k = 0$.

electron in the bottom level of the conduction band has only a small probability of emitting a photon and returning to the top of the valence band. This is of importance when selecting materials for some of the applications we are going to consider.

Transitions across the band gap are also responsible for the appearance of many solids. Because a solid is very concentrated, the probability that a photon whose energy corresponds to an allowed transition, is absorbed, is very high. Many such photons will therefore be absorbed at or near the surface of the solid. These photons will then be re-emitted in random directions so that some will be reflected back towards the source of radiation and some will travel further into the solid. Those travelling into the solid stand a very good chance of being re-absorbed and then re-emitted, again in random directions. The net effect is that the radiation does not penetrate the solid but is reflected by its surface. If the surface is sufficiently regular, then solids which reflect visible radiation appear shiny. Thus silicon whose band gap is at the lower end of the visible region and has allowed transitions covering most of the visible wavelengths, appears shiny and metallic. Many metals have strong transitions between the conduction band and a higher energy band which lead to their characteristic metallic sheen. Some metals such as tungsten and zinc have a band gap in the infrared and transitions in the visible are not so strong. These metals appear relatively dull. Gold and copper have strong absorption bands due to excitation of d band electrons to the s/p conduction band. In these elements, the d band is full and lies some way below the Fermi level (Figure 6.9). The reflectivity peaks in the yellow part of the spectrum and blue and green light are less strongly absorbed; hence the metals appear golden. Very thin films of gold appear green because the yellow and red light is absorbed and only the blue and green transmitted. Insulators typically have band gaps in the ultra-violet and unless there is a localized transition in the visible, appear colourless. The devices which we consider involving band gap transitions are, however, concerned with emission of light rather than absorption or reflection; the electrons being initially excited by electrical energy.

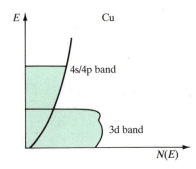

Figure 6.9 Band structure of copper.

6.3.1 Light-emitting diodes

Light-emitting diodes (LEDs) are effectively the reverse of photovoltaic cells (Chapter 2). In photovoltaic cells, light is used to produce an electrical voltage; in LEDs, a voltage is applied across a $p-n$ junction to produce light. Figure 6.10 shows a $p-n$ junction in a semiconductor such as GaAs.

The band structure shown is for the junction in the dark and with no electric field applied. Now suppose that an electrical field is applied so that the n-type is made negative relative to the p-type. Electrons will then flow from the n-type to the p-type. An electron in the conduction band moving to the p-type side can drop down into one of the vacancies in the valence band on the p-type side, emitting a photon in the process. This is more likely to happen if the transition is allowed, so that semiconductors with direct band gaps are usually used in such devices. To use the LED as a display, for example, it is then wired so that an electric field is applied across the parts making up the required letters or numbers. Different colours can be produced by using semiconductors of differing band gap. GaP produces red light, but by mixing in various proportions of aluminium to form $Ga_{1-x}Al_xP$, green or orange light can be produced.

It should be noted that semiconductors with indirect band gaps are used for LEDs, but in these cases, impurity levels play an important role. Thus GaP is used although it has an indirect band gap. Silicon, however, is not suitable because there is a non-radiative transition available to electrons at the bottom of the conduction band and these electrons donate thermal energy to the crystal lattice rather than emitting light when they return to the valence band.

When the electric field causes conduction electrons to move across the $p-n$ junction, the resulting situation is one in which the population in the

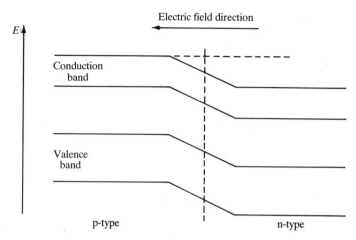

Figure 6.10 Energy bands near the junction in a *p–n* junction.

conduction band is greater than the thermal equilibrium population. An excess of electrons in an excited state is an essential feature of lasers and there are several semiconductor lasers based on the *p–n* junction. The best known of these is the GaAs laser.

6.3.2 The gallium arsenide laser

The GaAs laser actually contains a layer of GaAs sandwiched between layers of *p*- and *n*-type gallium aluminium arsenide ($Ga_{1-x}Al_xAs$). The band gap of gallium aluminium arsenide is larger than that of GaAs (Figure 6.11).

An electric field applied across the *p–n* junction as in LEDs, produces an excess of electrons in the conduction band of the GaAs. These electrons do not drift across into the gallium aluminium arsenide layer, however, because the bottom of the conduction band in this layer is higher in energy and the electrons would thus need to gain energy in order to move across. The excess conduction band electrons are therefore constrained to remain within the GaAs layer. Eventually one of these electrons drops down into the valence band, emitting a photon as it does so. This photon induces other conduction band electrons to return to the valence band and thus a coherent beam of light begins to build up. As in the ruby laser the initial burst of photons is reflected back by mirrors placed at the ends and induces more emission. Eventually a beam of infrared radiation is emitted. Several such lasers have been developed, most of them based on **III–V** compounds (that is compounds of In, Ga and/or Al with As, P or Sb). It is possible to manufacture materials with

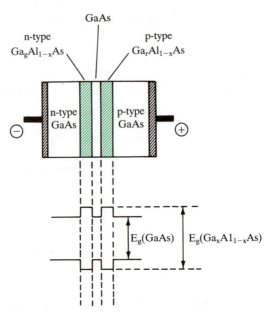

Figure 6.11 The arrangement of the different semiconductor regions in a GaAs laser. The band gap profile is shown below.

band gaps over the range 400–1300 nm by carefully controlling the ratios of the different elements.

Infrared semiconductor lasers are used, amongst many other things, for reading compact discs. A compact disc consists of a plastic disc coated with a highly reflective aluminium film and protected against mechanical damage by a layer of polymer. The original sound recording is split up into a number of frequency channels and the frequency of each channel given as a binary code (that is a series of 0s and 1s). This is converted to a series of pits in tracks on the disc, spaced approximately 1.6 μm apart. A laser is focused on the disc and reflected on to a photodetector. The pits cause some of the light to be scattered thus reducing the intensity of the reflected beam. The signal read by the photodetector is read as 1 when there is a high intensity of light and as 0 when the intensity is reduced by scattering. The binary code is thus recovered and can be converted back to sound.

Another use of semiconductor lasers is as the light source in fibre optics. For this purpose, long wavelength radiation well into the infrared is needed.

6.4 OPTICAL FIBRES

Optical fibres are used to transmit light in the way that metal wires are used to transmit electricity. For example, a telephone call can be sent along an optical fibre in the form of a series of light pulses from a laser. The intensity, time between pulses and length of pulse can be modified to convey the contents of the call in coded form. In order to transmit the information over useful distances (of the order of kilometres) the intensity of the light must be maintained so that there is still a detectable signal at the other end of the fibre. Thus much of the art of making commercial optical fibres lies in finding ways of reducing energy loss.

The first requirement is that the laser beam keeps within the fibre. Laser beams diverge less than conventional light beams so that using laser light is a help, but even so there is some tendency for the beam to stray outside the fibre. Therefore, fibres are usually constructed with a variable refractive index across the fibre. The beam is sent down a central core. The surrounding region has a lower refractive index than the core so that light deviating from a straight path is totally internally reflected and hence remains in the core (Figure 6.12).

More sophisticated designs vary the refractive index across the core. This is to keep signal pulses together as they travel along the fibre. In the simple case shown in Figure 6.12, the totally internally reflected rays travel a longer path than those that travel straight along the core. This will lead to a pulse being spread out in time. The refractive index is a measure of how fast light travels in a medium; the lower the refractive index, the faster the speed of light. If the outer parts of the core, therefore, have a lower refractive index, then the reflected light moves faster and this compensates for the longer pathlength.

There are bound to be some imperfections in the fibre and these are another source of energy loss. The imperfections cause scattering of the light of a type known as **Rayleigh scattering**. Rayleigh scattering does not cause any change in the wavelength of the light, only in its direction. The amount of scattering depends on $(1/\lambda^4)$, where λ is the wavelength and so there is much less scattering for longer wavelengths. Even the reduction

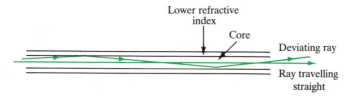

Figure 6.12 Rays of light travelling along an optical fibre.

in going from blue to red light is significant and is responsible for the colour of the sky. To reduce Rayleigh scattering, the lasers employed for optical fibre systems usually emit infrared radiation.

A third source of energy loss is absorption of light by the fibre. In a fibre several kilometres long, a very small amount of impurity can give rise to substantial absorption. On looking at a sheet of window glass edge-on, instead of being clear, the glass appears green. This is due to absorption by Fe^{2+} ions in the glass. A window pane is only about 50 cm across, so in a fibre of a few kilometres in length, there would be considerable loss due to such absorption. In a glass, the spectrum of the impurity ions is similar to that in a crystal, but because the ions occupy several different types of site in a glass, the absorption bands are wider, each site giving rise to a band at a slightly different wavelength. In an optical fibre 3 km long operating at 1300 nm in the near infrared, the intensity of the Fe^{2+} absorption is still such that a concentration of 2 parts in 10^{10} would reduce the amount of radiation by one-half. Materials for optical fibres must therefore be very pure. One reason why silica has been widely used is because high purity silicon tetrachloride, developed for the semiconductor industry, is commercially available as a starting material.

Metal ions are not the only source of absorption, however. Using infrared radiation means there is likely to be loss due to absorption by molecular vibrations. In silica glass the structure may contain dangling SiO bonds which easily react with water to form OH bonds. The vibrational frequencies of OH bonds are high and close to the frequencies used for transmission. It is important, therefore, to exclude water when manufacturing silica optical fibres. Even when water is excluded and there are no OH bonds, absorption by vibrational modes cannot be neglected. SiO bonds vibrate at lower frequencies than OH bonds, and so the maximum in the absorption does not interfere. However, the SiO absorption is very strong and the peak tails into the region of the transmission frequency. There has been some research into substances with lower vibrational frequencies than silica, particularly fluorides, but as yet such substances are not economically viable, being difficult to manufacture and more expensive than silica.

There are still losses in the fibres developed for commercial use, nonetheless these have been reduced to a point where transmission over several kilometres is possible. As well as transmission of information in telephone systems and similar applications, it has been suggested that optical devices may replace conventional electronics in more advanced applications such as computers. For such applications, it will be necessary to develop optical switches, amplifiers and so on. We conclude this chapter by looking at one such device.

6.4.1 Optical switches

The refractive index of a crystal is not always a fixed quantity, but can be made to vary. In some materials, an intense beam of light such as produced by a laser can itself alter the refractive index. Another way to alter the value is to apply an electrical field, and this is the basis of a commercially available optical switch.

The lithium niobate switch consists of a crystal of lithium niobate ($LiNbO_3$) with two optical channels formed in it by diffusing in titanium (Figure 6.13). The titanium alters the refractive index and the radiation is contained within the channels in the same way as radiation is confined in optical fibres. That is, slightly deviating rays are totally internally reflected due to the different refractive index outside the channels. Applying an electric field alters the refractive index between the channels so that the radiation is no longer reflected and can switch from one channel to the other. To see why the refractive index is altered by composition and by applying an electric field, the atomic origin of the refractive index is considered briefly.

Electromagnetic radiation has associated with it an oscillating electric field. Even when the radiation is not absorbed, this field has an effect on the electrons in the solid. Think of an electrical field applied to an atom, and imagine the electrons pulled by the field so that the atom is no longer spherical. The applied field produces a separation of the centres of positive and negative charge, i.e. it induces an electric dipole moment. (A molecule in a solid may also have a permanent electric dipole moment

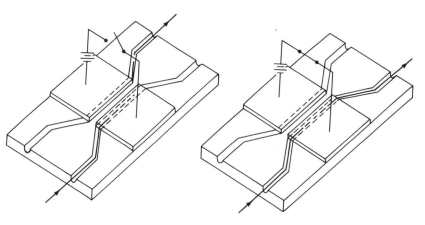

Figure 6.13 An electro-optical switch.

produced by an unequal distribution of bonding electrons between the nuclei but this is present in the absence of an applied field.)

The oscillating field of the radiation can be thought of as pulling the electrons alternately one way and then the other. The amount the electrons respond depends on how tightly bound the electrons are to the nucleus. This property is called the **polarizability** and is higher for large ions with low charge, e.g. Cs^+, than for small, highly charged ions such as Al^{3+}. If there is a high concentration of polarizable ions in a solid, then the radiation will be slowed down, i.e. its refractive index will increase. Lithium niobate contains Li^+ and (formally) 'Nb^{5+}' which will have low polarizabilities; adding titanium (which formally goes into the solid as 'Ti^{4+}') increases the refractive index because the titanium ions have a slightly higher polarizability. Applying an electric field across the solid distorts the electron distribution and causes the refractive index to increase.

Adjusting the refractive index by adding carefully selected impurities is also useful in other applications. For example, controlling the refractive index of glass is very important when making lenses for telescopes, binoculars and cameras. Lead ions, Pb^{2+}, are highly polarizable and are used to produce glass of high refractive index, such as decorative crystal.

We have used the lithium niobate switch as an example of a currently available optical device for use in information storage and transmission. Others are or will no doubt soon be available. Examples include computer storage discs based on technology developed for compact discs, or an optical switch in which the intensity of the light itself is used to transfer radiation between paths. Many current devices, however, depend on electronic properties or on magnetic properties. Electronic properties were discussed in Chapter 2. Chapter 7 considers the nature and applications of magnetic properties.

FURTHER READING

West, A.R. (1988) *Basic Solid State Chemistry*, Chapter 8, John Wiley, New York.
  See comments following Chapter 2. The last part of chapter 8 deals with applications of luminescence and lasers.
Duffy, J.A. (1990) *Bonding, Energy Levels and Bands in Inorganic Solids*, Chapters 2, 3, 5 and 8, Longman, London.
  See comments after Chapter 2. This book is of most use for optical properties, although much of it is written at a level more suitable for 3rd year undergraduates and post graduates.
Moore, W.J. (1967) *Seven Solid States*, Chapter VI, W.A. Benjamin Inc.
  See comments after Chapter 2. The description of the energy levels and the fundamental laser process are useful.

Nassau, K. (1983) *The Physics and Chemistry of Color*, Chapters 5, 8 and 14, New York.
See comments after Chapter 3.

The following three articles are very readable and at a suitable level.
Hill, C.G.A. (1983) Inorganic luminescent materials, *Chemistry in Britain*, September, 723.
Goodman, C.H.L. (1983) Optical fibres, *Chemistry in Britain*, September, 745.
Deluca, J.A. (1980) An introduction to luminescence in inorganic solids, *J. Chem. Ed.*, **57**, 541.

QUESTIONS

1. In the oxide MnO, Mn^{2+} ions occupy octahedral holes in an oxide lattice. The degeneracy of the $3d$ levels of manganese are split into two, as for Ti^{3+}. The five d electrons of the Mn^{2+} ions occupy separate d orbitals and have parallel spins. Explain why the absorption lines due to transitions between the two $3d$ levels are very weak for Mn^{2+}.

2. Figure 6.14 shows the energy levels of Nd^{3+} in yttrium aluminium garnet ($Y_3Al_5O_{12}$) which are involved in the laser action of this crystal (known as the Nd YAG laser). Describe the processes that occur when the laser is working.

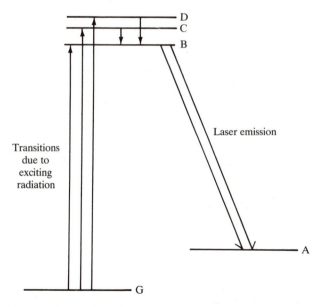

Figure 6.14 Energy levels of Nd^{3+} in yttrium aluminium garnet.

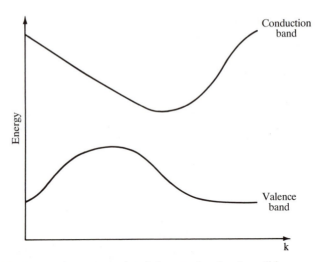

Figure 6.15 A plot of energy against k for two bands of a solid.

3. A phosphor commonly used on television screens is ZnS doped with Cu^+. This is more efficient at transferring energy to the impurity sites for emission than are the phosphors based on phosphates as host. ZnS is a semiconductor. Suggest a reason for the efficiency of transfer in this solid.

4. In Chapter 3, the photographic process was described. Outline the processes leading to the formation of a silver atom when red light is absorbed by a grain of AgBr in an emulsion containing a sensitizer. AgBr has an indirect band gap. Explain why this is useful. Compare the processes you describe with those in phosphors.

5. Figure 6.15 shows two bands for a semiconductor. Is the band gap of this solid direct or indirect?

6. Explain why silicon is used for solar cells but not for LEDs.

7. The core of silica optical fibres contains some B_2O_3, GeO_2 and P_2O_5. How do these impurities increase the refractive index?

ANSWERS

1. In Mn^{2+}, an electron can only go from one $3d$ level to another if it changes its spin. Transitions in which an electron changes its spin are forbidden and so give rise to weak spectral lines.

2. Radiation from a lamp excites electrons from the ground state, G, to states B, C, D. Electrons in states C and D undergo non-radiative transitions to state B. Radiative transitions from B to lower states are forbidden and so a large population of electrons in state B builds up. Eventually a photon is emitted and an electron goes to state A. This photon induces further emission from B to A, eventually producing a beam of laser light. Transitions from B to both A and G must be forbidden for a large population to build up in B, but the transition B → A must be less forbidden than the transition B → G as it is B → A that gives rise to laser action, and hence must be the more likely to occur.

3. ZnS is a semiconductor with a full valence band and empty conduction band. When the phosphor is illuminated, electrons are promoted to the conduction band. Since the orbitals in this band are delocalized, the energy can easily be transferred to other parts of the crystal, in particular to the dopant atoms.

4. The sensitizer acts like an impurity. It absorbs the radiation and goes to an excited state. The excited electron undergoes a radiationless transition to a level in the conduction band of AgBr. Since it is now in a delocalized level, the electron can travel through the crystal until it combines with an interstitial Ag^+ ion to form a silver atom. The indirect band gap helps by reducing the probability of the electron in the conduction band dropping into a vacancy in the valence band. The process is similar to that of phosphors. However, with phosphors it is the light emitted by the impurity that is important whereas for photography it is the light absorbed by the sensitizer that matters.

5. The solid has an indirect band gap because the lowest energy interband transition (that is from the highest energy level in the valence band to the lowest energy level in the conduction band) does not correspond to $\Delta k = 0$ and is therefore forbidden.

6. Silicon is an indirect band gap solid with an available non-radiative pathway from the conduction band to the valence band. In photovoltaic cells, electrons are promoted from the valence band to the conduction band and are then used to do electrical work. The promoted electrons do not return directly to the valence band either by emitting energy or by a non-radiative pathway. In LEDs it is the return of the electrons to the valence band by emitting light that is important. This return has low probability because of the indirect band gap and the electrons use the non-radiative pathway instead. Promotion to the

conduction band in the solar cell will also be of low probability but there is no competing non-radiative route.

7. These oxides (at least formally) contain small highly charged ions. These have low polarizability and hence impart high refractive indices.

Magnetic properties of solids | 7

7.1 INTRODUCTION

Weak magnetic effects occur in all substances, gas and liquid as well as solid, but the greater proximity of the atoms in a solid can lead to stronger cooperative effects. If the interaction is such that the magnetism on the atoms is aligned, this leads to the very strong magnetic effect known as **ferromagnetism**. There are other effects though that lead to a cancelling (**antiferromagnetism**) or partial cancelling (**ferrimagnetism**) of the magnetism of different atoms. Ferro- and ferrimagnetism have many commercial applications from compass needles and watch magnets to audio- and videotapes and computer memory devices. This chapter will cover mainly ferro-, ferri-, and antiferromagnetism and some of these applications.

First we compare these types of magnetism with the weak effects due to isolated atoms or molecules. The main weak magnetic effects are **diamagnetism** and **paramagnetism**. These are distinguished by the sign of the magnetic susceptibility; diamagnetic substances having negative susceptibility and being repelled by an applied magnetic field whilst paramagnetic substances have positive susceptibilities and are attracted by an applied field.

7.2 MAGNETIC SUSCEPTIBILITY

A magnetic field produces lines of force which penetrate the medium the field is applied to. These lines of force show up for example when you scatter iron filings on a piece of paper covering a bar magnet. The density of these lines of force is known as the **magnetic flux density**, B. In a vacuum, the magnetic field, H, and the magnetic flux density are related by the permeability of free space, μ_0.

$$B = \mu_0 H. \tag{7.1}$$

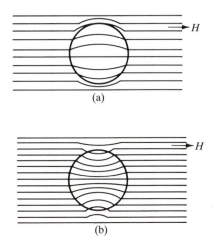

Figure 7.1 Flux density in (a) a diamagnetic and (b) a paramagnetic sample.

If a magnetic material is placed in the field, however, it can increase or decrease the flux density. Diamagnetic materials reduce the density of the lines of force as shown in Figure 7.1. Paramagnetic materials increase the flux density.

The field of the sample in the applied field is known as its **magnetization (M)**. The magnetic flux density is given by equation (7.2).

$$B = \mu_0(H + M). \tag{7.2}$$

The magnetization is usually discussed in terms of the **magnetic susceptibility, χ**, where $\chi, = M/H$.

Diamagnetism is present in all substances but is very weak so that it is not normally observed if other effects are present. It is produced by circulation of electrons in an atom or molecule. Atoms or molecules with closed shells of electrons are diamagnetic. Unpaired electrons, however, give rise to paramagnetism. Simple paramagnetic behaviour is found for substances such as liquid oxygen or transition metal complexes in which the unpaired electrons on different centres are isolated from each other. In a magnetic field the magnetic moments on different centres tend to align with the field and hence with each other, but this is opposed by the randomizing effect of thermal energy and in the absence of a field, the unpaired electrons on different centres are aligned randomly. The interplay of applied field and thermal randomization leads to the temperature dependence described by the **Curie law**

$$\chi = C/T, \tag{7.3}$$

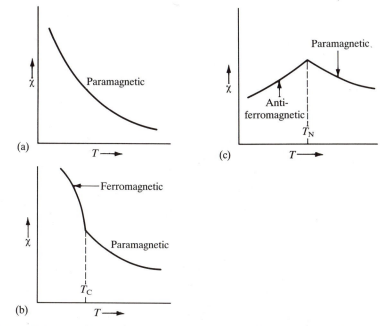

Figure 7.2 Variation of magnetic susceptibility with temperature for (a) a para-magnetic substance, (b) a ferromagnetic substance and (c) an antiferromagnetic substance.

where χ is the magnetic susceptibility, C is a constant known as the Curie constant and T is the temperature in Kelvin.

Different temperature dependence is observed when there is co-operative behaviour. The changeover from independent to cooperative behaviour is associated with a characteristic temperature. For ferro-magnetism, the Curie law becomes

$$\chi = C/(T - T_c), \tag{7.4}$$

where T_c is the Curie temperature. For antiferromagnetism, the tem-perature dependence is of the form

$$\chi = C/(T + T_N), \tag{7.5}$$

where T_N is the Néel temperature. These two behaviours are illustrated in Figure 7.2.

Ferrimagnetism has a more complicated form of temperature depen-dence with ions on different sites having different characteristic tem-peratures. The characteristics of the various types of magnetism that we shall be concerned with are given in Table 7.1.

Table 7.1 Characteristics of the types of magnetism

Type	Sign of χ	Typical χ value (calculated using SI units)	Dependence of χ on H	Change of χ with increasing temperature	Origin
Diamagnetism	−	$-(1-600) \times 10^{-5}$	Independent	None	Electron charge
Paramagnetism	+	0–0.1	Independent	Decreases	Spin and orbital motion of electrons on individual atoms
Ferromagnetism	+	$0.1-10^7$	Dependent	Decreases	Cooperative interaction between magnetic moments of individual atoms
Antiferromagnetism	+	0–0.1	May be dependent	Increases	
Pauli paramagnetism	+	10^{-5}	Independent	None	Spin and orbital motion of delocalized electrons

7.3 PARAMAGNETISM IN METAL COMPLEXES

In solids containing metal complexes such that the unpaired electrons on the different metal atoms are effectively isolated, the susceptibility can be discussed in terms of magnetic moments. The isolated metal complex can be thought of as a small magnet. Each complex in a solid will produce its own magnetic field due to the unpaired electrons. If the solid consists of one type of complex, then each complex produces the same magnitude magnetic field. However, thermal motion causes the orientation of these fields to be random. In the Curie law (7.3), the temperature dependence is the result of this thermal motion but the constant, C, gives us information on the value of the magnetic field, known as the **magnetic moment**, μ, of the complex. The magnetic moment, unlike the susceptibility, does not normally vary with temperature. The dimensionless quantity χ in equation (7.2) is the susceptibility per unit volume. To obtain the size of the magnetic field due to an individual complex, χ is divided by the specific gravity to give the susceptibility per unit mass of the sample and then multiplied by the relative molecular mass to obtain the molar susceptibility χ_m. Assuming that each complex has a fixed magnetic moment, μ, and that the orientation is randomized by thermal motion, it can be shown that χ_m is proportional to μ^2 as in equation (7.6).

$$\chi_m = \frac{N_A \mu_0}{3kT} \mu^2, \tag{7.6}$$

where N_A is Avogadro's number, k is Boltzmann's constant, μ_0 is the permeability of free space and T is the temperature in Kelvin. In SI units, χ_m is in $m^3 \, mol^{-1}$ and the magnetic moment in joules per tesla ($J\,T^{-1}$). It is usual to quote μ in Bohr magnetons (BM or μ_B) where one Bohr magneton has a value of $9.274 \times 10^{-24} \, J\,T^{-1}$.

The magnetic moment, μ, is a consequence of the angular momentum of the unpaired electrons. The electrons possess both spin and orbital angular momentum. For the first row transition elements, the contribution from the orbital angular momentum is greatly reduced or 'quenched' as a consequence of the lifting of the five-fold degeneracy of the $3d$ orbitals. In complexes of these atoms, the magnetic moment is often close to that predicted for spin angular moment only:

$$\mu_s = g\sqrt{S(S + 1)}, \tag{7.7}$$

where g is a constant which for a free electron has the value 2.00023, and μ_s is in Bohr magnetons. The value of S depends on the number of unpaired electrons and Table 7.2 gives the values of S and μ_s for the possible numbers of unpaired $3d$ electrons. Contributions from orbital angular momentum cause deviations from these values.

For complexes containing heavier metal ions, interaction of spin and

Table 7.2 Values of S and μ_s for unpaired $3d$ electrons

No. of unpaired electrons	Spin quantum number (S)	Magnetic moment μ_S in Bohr magnetons
1	$\frac{1}{2}$	1.73
2	1	2.83
3	$\frac{3}{2}$	3.87
4	2	4.90
5	$\frac{5}{2}$	5.92

orbital angular momenta is greater. For the lanthanides the magnetic moment depends on the total angular momentum of the electrons, $\mathbf{J}$, not just or mainly on the spin angular momentum. The total angular momentum, $\mathbf{J}$, is the vector sum of the orbital, $\mathbf{L}$, and spin, $\mathbf{S}$, angular momenta.

$$\mathbf{J} = \mathbf{L} + \mathbf{S}. \tag{7.8}$$

If $\mathbf{J}$ is the quantum number for the total electronic angular momentum, then the magnetic moment is given by:

$$\mu = g\sqrt{J(J + 1)}, \tag{7.9}$$

where

$$g = 1 + \frac{J(J + 1) + S(S + 1) - L(L + 1)}{2J(J + 1)}.$$

Table 7.3 gives values of the magnetic moment for different numbers of f electrons, calculated from equation (7.9). Thus the magnetic moment gives information on the number of unpaired electrons per atom and whether the orbital angular momentum contributes.

7.4 FERROMAGNETIC METALS

When discussing the electrical conductivity of metals, we described them in terms of ionic cores and delocalized valence electrons. The core electrons contribute a diamagnetic term to the magnetic susceptibility, but the valence electrons can give rise to paramagnetism or one of the cooperative effects we have described.

In filling the conduction band, we have implicitly put electrons into energy levels with paired spins. Even in the ground state of simple molecules such as O_2, however, it can be more favourable to have electrons in different orbitals with parallel spins than in the same orbital

Table 7.3 Values of the magnetic moment for different numbers of f electrons

No. of f electrons	No. of unpaired electrons	J	L	S	μ in Bohr magnetons
1	1	$\frac{5}{2}$	3	$\frac{1}{2}$	2.54
2	2	4	5	1	3.58
3	3	$\frac{9}{2}$	6	$\frac{3}{2}$	3.62
4	4	4	6	2	2.68
5	5	$\frac{5}{2}$	5	$\frac{5}{2}$	0.85
6	6	0	3	3	0
7	7	$\frac{7}{2}$	0	$\frac{7}{2}$	7.94
8	6	6	3	3	9.72
9	5	$\frac{15}{2}$	5	$\frac{5}{2}$	10.65
10	4	8	6	2	10.61
11	3	$\frac{15}{2}$	6	$\frac{3}{2}$	9.58
12	2	6	5	1	7.56
13	1	$\frac{7}{2}$	3	$\frac{1}{2}$	4.54

with paired spins. This occurs when there are degenerate or nearly degenerate levels. In an energy band, there are many degenerate levels and many levels very close in energy to the highest occupied level. It might well be favourable then to reduce electron repulsion by having electrons with parallel spin singly occupying levels near the Fermi level. To obtain a measurable effect, however, the number of parallel spins would have to be comparable with the number of atoms; 10^3 unpaired spins would not be noticed in a sample of 10^{23} atoms. Unless the density of states is very high near the Fermi level, a large number of electrons would have to be promoted to high energy levels in the band in order to achieve a measurable number of unpaired spins. The resulting promotion energy would be too great to be compensated for by the loss in electron repulsion. In the wide bands of the simple metals, the density of states is comparatively low, so that in the absence of a magnetic field, few electrons are promoted.

When a magnetic field is applied, the electrons will acquire an extra energy term due to interaction of their spins with the field. If the spin is parallel to the field, then its magnetic energy is negative, i.e. the electrons are at lower energy than they were in the absence of a field. For an electron with spin antiparallel to the field, it is now worthwhile to go to a higher energy state and change spin so long as the promotion energy is not more than the gain in magnetic energy. This will produce a measurable imbalance of electron spins aligned with and against the field and hence the solid will exhibit paramagnetism. This type of paramagnetism is known as **Pauli paramagnetism** and is a very weak effect giving a magnetic susceptibility much less than that due to isolated spins and comparable in magnitude to diamagnetism.

For a very few metals, however, the unpaired electrons in the conduction band can lead to ferromagnetism. In the whole of the periodic table, only iron, cobalt, nickel and a few of the lanthanides (Gd, Tb) possess this property. So, what is so special about these elements that confers this uniqueness on them? It is not their crystal structure; they each have different structures and the structures are similar to those of other non-ferromagnetic metals. Iron, cobalt and nickel, however, do all have a nearly full, narrow $3d$ band.

The $3d$ orbitals are less diffuse than the $4s$ and $4p$, i.e. they are concentrated nearer the atomic nuclei. This leads to less overlap so that the $3d$ band is a lot narrower than the $4s/4p$ band. Furthermore, there are five $3d$ orbitals so that for a crystal of N atoms, $5N$ levels must be accommodated. With more electrons and a narrower band, the average density of states must be much higher than in ns/np bands. In particular the density of states near the Fermi level is high. In this case, it is energetically favourable to have substantial numbers of unpaired electrons at the cost of populating higher energy levels. Thus these elements have large numbers of unpaired electrons even in the absence of a magnetic field. For iron, for example, in a crystal of N atoms there are up to $2.2N$ unpaired electrons all with their spins aligned parallel. Note the contrast with a paramagnetic solid containing transition metal complex ions where each ion may have as many as five unpaired electrons but in the absence of a magnetic field, electrons on different ions are aligned randomly.

Ferromagnetism thus arises from the alignment of electron spins throughout the solid, and this occurs for partially filled bands with a high density of states near the Fermi level. $4d$ and $5d$ orbitals are more diffuse than $3d$ and produce wider bands so that ferromagnetism is not observed in the second and third row transition elements. The $3d$ orbitals themselves become less diffuse across the transition series and lower in energy. In titanium the valence electrons are in the $4s/4p$ band with low density of states and, at the other end of the row in copper, the $3d$ band has dropped in energy so that the Fermi level is in the $4s/4p$ band. Thus it is only at the middle of the series that the Fermi level is in a region of high density of states. Schematic band diagrams for Ti, Ni and Cu are given in Figure 7.3. The occupied levels are indicated by shading.

The pure elements are not always suitable for applications requiring a metallic ferromagnet and many ferromagnetic alloys have been produced. Some of these contain one or more ferromagnetic elements and among these alloys of iron, cobalt and nickel with the lanthanides, e.g. $SmCo_5$, $Nd_2Fe_{14}B$, have produced some of the most powerful permanent magnets

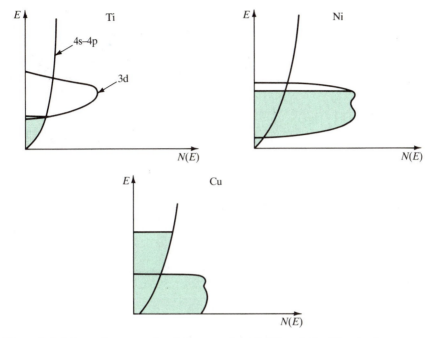

Figure 7.3 Schematic energy level diagrams for Ti, Ni and Cu. The shading represents occupied levels.

known. In the lanthanide alloys, f electrons contribute to the magnetism. Potentially this could lead to a very high magnetization because there are seven f orbitals and so a maximum possible magnetization corresponding to seven electrons per atom. The theoretical maximum magnetization for the transition metals is five electrons per atom, as there are only five d orbitals. In practice this maximum is never reached. In the pure lanthanide metals, the overlap of f orbitals is so small that they can be regarded as localized. In the ferromagnetic lanthanides, the magnetism is produced by delocalized d electrons. The interaction between these d electrons and the localized f electrons causes alignment of the d and f electrons in order to reduce electron repulsion. Thus f electrons on different atoms are aligned through the intermediary of the d electrons. In alloys, the f electrons can align via the transition metal d electrons, and although not all the d and f electrons are aligned, it can be seen that high values of the magnetization could be achieved. It is not surprising, then, that it is these transition metal/lanthanide alloys that are the most powerful magnets. Other alloys can be made from non-magnetic elements such as manganese and in these the overlap of d orbitals is brought into the range necessary for ferromagnetism by altering the interatomic distance from that in the element.

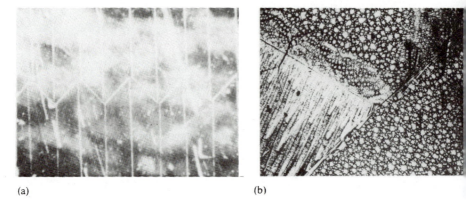

(a)   (b)

Figure 7.4 (a) Domain patterns for a single crystal of iron containing 3.8% of silicon. The white lines show the boundaries between the domains. From R. Eisberg and R. Resnick (1985) *Quantum Physics of Atoms, Molecules, Solids, Nuclei and Particles*, © John Wiley (courtesy of H.J. Williams, Bell Telephone Laboratories). (b) Magnetic domain patterns on surface of an individual crystal of iron. From W.J. Moore, (1967) *Seven Solid States*, W.A. Benjamin Inc., © W.J. Moore.

The usefulness of a particular ferromagnetic substance depends on factors such as the size of the magnetization produced, how easily the solid can be magnetized and demagnetized and how readily it responds to an applied field. The number of unpaired electrons will determine the maximum field, but the other factors depend on the structure of the solid and the impurities it contains.

7.4.1 Ferromagnetic domains

A drawback to the explanation above may have occurred to you. If $2.2N$ electrons are all aligned in any sample of iron, why are not all pieces of iron magnetic? The reason for this is that our picture only holds for small volumes of metal (about $10^{-14}\,\mathrm{m}^3$) within a crystal, called **domains**. Within each domain the spins are all aligned, but the different domains are aligned randomly with respect to each other. It is actually possible to see these domains through a microscope by spreading finely divided iron powder on the polished surface of a crystal (Figure 7.4). What then causes these domains to form?

The spins tend to align parallel because of short range **exchange interactions** stemming from electron–electron repulsion, but there is also a longer range **magnetic dipole interaction** tending to align the spins antiparallel. Consider building up a domain starting with just a few spins, initially the exchange interactions dominate and so the spins all lie parallel. As more spins are added, an individual spin will be subjected

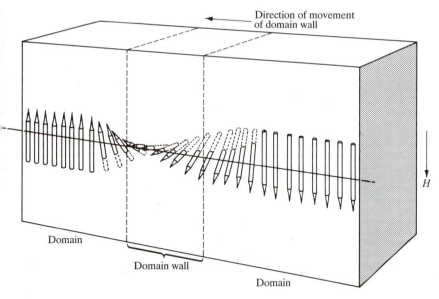

Figure 7.5 Movement of a domain wall. The broken lines show the domain wall, which separates two domains with magnetic moments lined up in opposite directions. The moments twist to align with the applied field, H, and the wall moves in the direction of the arrow.

to a greater and greater magnetic dipole interaction. Eventually the magnetic dipole interaction overcomes the exchange interaction and the adjacent piece of crystal has the spins aligned antiparallel to the original domain. Thus within domains exchange forces keep the spins parallel, whereas the magnetic dipole interaction keeps the spins of different domains aligned in different directions.

When a magnetic field is applied to a ferromagnetic sample, the domains all tend to line up with the field. The alignment can be accomplished in two ways. Firstly a domain of correct alignment can grow at the expense of a neighbouring domain. Between the two domains is an area of finite thickness known as the domain wall. Changeover from the alignment of one domain to that of the next is gradual within the wall. When the magnetic field is applied, the spins in the wall nearest the aligned domain alter their spins to line up with the bulk of the domain. This causes the next spins to alter their alignment. The net effect is to move the wall of the domain further out (Figure 7.5). This process is reversible; the spins return to their former state after the magnetic field is removed.

If impurities or defects are present, it becomes harder for a domain to grow; there is an activation energy to aligning the spins through the

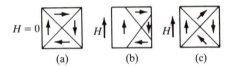

Figure 7.6 Magnetization processes according to the domain model: (a) un-magnetized; (b) magnetized by domain growth (boundary displacement); (c) magnetized by domain rotation (spin alignment).

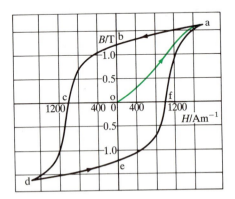

Figure 7.7 B–H curve for a typical hard steel.

defect and so a larger magnetic field is required. Once the domain has grown past the defect, however, it cannot shrink back once the magnetic field is removed because this will also need an energy input. In this case the solid retains magnetization. The amount retained depends on the number and type of defects. Thus steel (which is iron with a high impurity content) remains magnetic after the field is removed whereas soft iron (which is much purer) retains hardly any magnetization.

The second mechanism of alignment, which only occurs in strong magnetic fields, is that the interaction of the spins with the applied field becomes large enough to overcome the dipole interaction and entire domains of spins change their alignment simultaneously. The two mechanisms are compared in Figure 7.6.

The magnetic behaviour of different ferromagnetic substances is shown by their **hysteresis curves**. This is a plot of magnetic flux density, B, against applied magnetic field, H. If we start with a non-magnetic sample in which all the domains are randomly aligned, then in the absence of a magnetic field, B and H are zero. As the field is increased, the flux density also increases. The plot of B against H is shown in Figure 7.7. Initially the curve is like 'oa' which is not simply a straight line because the magnetization is increasing with the field. At point 'a' the magnetization has reached its maximum value; all the spins in the sample are aligned. When the applied field is reduced, the flux density does not follow the initial curve. This is due to the difficulty of reversing processes where domains have grown through crystal imperfections. A sufficiently large field in the reverse direction to provide the activation energy for realignment through the imperfection must be applied before the magnetization process can be reversed. At point b therefore where H is zero, B is not zero because there is still a contribution from the magnetization, M. The magnetization at this point is known as the **remanent magnetization**. The field that needs to be applied in the reverse direction to reduce the magnetization to zero is the **coercive force** and is equal to the distance 'oc'.

7.4.2 Permanent magnets

Substances used as permanent magnets need a large coercive force, so that they are not easily demagnetized, and preferably should have a large remanent magnetization. These substances have fat hysteresis curves. They are often made from alloys of iron, cobalt or nickel which form with small crystals and include non-magnetic areas so that domain growth and shrinkage are difficult. Magnets for electronic watches for example are made from samarium/cobalt alloys. The best known of these alloys is $SmCo_5$ which has a coercive force of $6 \times 10^5 \, A \, m^{-1}$ compared to $50 \, A \, m^{-1}$ for pure iron.

7.5 FERROMAGNETIC COMPOUNDS – CHROMIUM DIOXIDE

Chromium dioxide (CrO_2) crystallizes with a rutile structure and is ferromagnetic with a Curie temperature of 392 K. Like VO and TiO, CrO_2 has metal $3d$ orbitals which can overlap to form a band. In CrO_2, however, this band is very narrow and so like iron, cobalt and nickel, CrO_2 displays ferromagnetism. The dioxides later in the row have localized $3d$ electrons (e.g. MnO_2) and are insulators or semiconductors. TiO_2 has no $3d$ electrons and is an insulator. VO_2 has a different structure at room temperature and is a semiconductor. It does, however, undergo a phase transition to a metal at 340 K, when it becomes Pauli paramagnetic. So CrO_2 occupies a unique position amongst the dioxides, similar to that of iron, cobalt and nickel amongst the first row transition metals, in which dioxides of elements to the left have wide bands of delocalized electrons and elements to the right have dioxides with localized $3d$ electrons. Because the metal atoms are further apart in the dioxides than in the elemental metals, the narrow bands that give rise to ferromagnetism occur earlier in the row than for the metallic elements.

7.5.1 Audiotapes

Commercial interest in CrO_2 lies in its use as a magnetic powder on audiotapes. Recording tape usually consists of a polyester tape impregnated with needle-like crystals of a magnetic material such as CrO_2 or γ-Fe_2O_3. The recording head has an iron core with a coil wrapped round it and a gap where the tape passes across it. Sound waves from the voice or music to be recorded hit a diaphragm in the microphone. The vibrations of the diaphragm cause a linked coil of wire to move in and out of a magnetic field. This causes a fluctuating electric current in the coil; the current depends on the frequency of the motion of the coil and hence on the frequency of the sound waves. The varying current is passed to the coil on the recording head, producing a varying magnetic field in the (soft) iron core. This in turn magnetizes the particles on the tape and the strength and direction of this magnetization is a record of the original sound. To play the tape, the whole process is reversed.

Materials for recording tape therefore need to retain their magnetization so that the recording is not accidentally erased. Having needle-like crystals aligned with the recording field helps here, but the material should also have a high coercive force. Chromium dioxide fulfils these requirements and has a high magnetization giving a large range of response and thus a high quality of reproduction. It does have some drawbacks, however; it has a relatively low Curie temperature, so that recordings can be erased by heating, and it is toxic.

7.6 ANTIFERROMAGNETISM – TRANSITION METAL MONOXIDES

These oxides have been discussed in Chapter 2. They are all based on the sodium chloride structure but have varying electrical properties. In this section we shall see that their magnetic properties are equally varied. In TiO and VO, the $3d$ orbitals are diffuse and form delocalized bands. These oxides are metallic conductors. The delocalized nature of the $3d$ electrons also determines the magnetic nature of these salts and, like the simple metals, they are Pauli paramagnetic. MnO, FeO, CoO and NiO have localized $3d$ electrons and are paramagnetic at high temperatures. On cooling, however, the oxides become antiferromagnetic. The Néel temperatures for this transition are 122, 198, 293 and 523 K, respectively.

In **antiferromagnetism**, the spins on different nuclei interact cooperatively but in such a way as to cancel out the magnetic moments. Antiferromagnetic materials therefore show a drop in magnetic susceptibility at the onset of cooperative behaviour. The temperature which characterizes this process is known as the **Néel temperature** (Figure 7.2c).

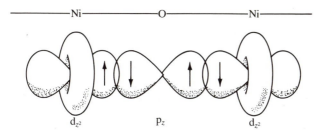

Figure 7.8 Overlap between Ni d_{z^2} orbitals and O p_z orbitals in NiO.

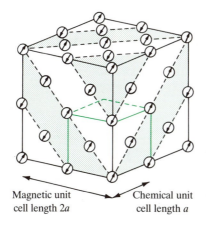

Magnetic unit
cell length $2a$

Chemical unit
cell length a

Figure 7.9 Magnetic unit cell of NiO, with the crystallographic or chemical unit cell indicated in colour.

The appearance of cooperative behaviour suggests that d electrons on different ions interact but the electronic properties are explained by assuming that the d electrons are localized. So how do we reconcile these two sets of properties?

The magnetic interaction in these compounds is thought to arise indirectly through the oxide ions; a mechanism known as **superexchange**. In a crystal of, say, NiO there is a linear Ni–O–Ni arrangement. The d_{z^2} orbital on the nickel can overlap with the $2p_z$ on oxygen, leading to partial covalency. The incipient NiO bond will have the d_{z^2} electron and a $2p_z$ electron paired. The oxide ion has a closed shell and so there is another $2p_z$ electron which must have opposite spin. This electron forms a partial bond with the next nickel and so the d_{z^2} on this nickel pairs with the $2p$ electron of opposite spin. The net result is that adjacent nickel ions have opposed spins (Figure 7.8).

The alternating spin magnetic moments in antiferromagnets such as

NiO can be observed experimentally using neutron diffraction. Because neutrons have a magnetic moment, a neutron beam used for diffraction responds not only to the nuclear positions but also to the magnetic moments of the atoms. X-rays, on the other hand, have no magnetic moment and respond to electron density and hence to atomic positions. The structure of NiO as found by X-ray diffraction is a simple NaCl structure. When the structure is determined by neutron diffraction, however, extra peaks appear which can be interpreted as giving a magnetic unit cell twice the size of the X-ray-determined unit cell. The positions of the nickel ions in this cell are shown in Figure 7.9. The normal crystallographic unit cell is bounded by identical atoms. The magnetic unit cell is bounded by identical atoms with identical spin alignment. The shading indicates layers of nickel ions parallel to the body diagonal of the cube. The spins of all nickel ions in a given layer are aligned parallel but antiparallel to the next layer.

7.7 FERRIMAGNETISM – FERRITES

The name ferrite was originally given to a class of mixed oxides having an **inverse spinel structure** and the formula AFe_2O_4, where A is a divalent metal ion. The term is sometimes extended to include other oxides, not necessarily containing iron, which have similar magnetic properties. In this section we shall concentrate on the original group.

The spinel structure is a common mixed oxide structure, typified by spinel itself, $MgAl_2O_4$, in which the oxide ions are in a face-centred cubic close-packed array. For an array of N oxide ions there are N octahedral

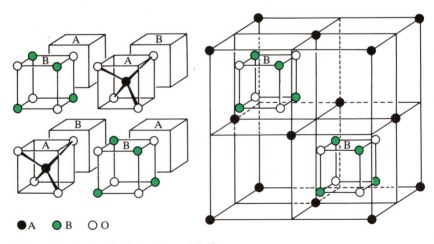

●A   ●B   ○O

Figure 7.10 The spinel structure, AB_2O_4.

holes (Al^{3+}) and the trivalent ions occupy half of the octahedral sites (Figure 7.10). In addition, there are $2N$ tetrahedral sites and the divalent ions (Mg^{2+}) occupy one-eighth of these. In the inverse spinel structure, the oxide ions are also in a cubic close-packed arrangement, but the divalent metal ions occupy octahedral sites and the trivalent ions are equally divided amongst tetrahedral and octahedral sites.

Ions on octahedral sites interact directly with each other and their spins align parallel. The ions on octahedral sites also interact with those on tetrahedral sites but in this case, they interact through the oxide ions and the spins align antiparallel as in NiO.

In ferrites, the Fe^{3+} ions on octahedral sites are aligned antiparallel to those on tetrahedral sites, so that there is no net magnetization from these ions. The divalent A ions, however, if they have unpaired electrons, tend to align their spins parallel with those of Fe^{3+} on adjacent octahedral sites, and hence with those of other A^{2+} ions. This produces a net ferromagnetic interaction for ferrites in which A^{2+} has unpaired electrons. The magnetic structure of a ferrimagnetic ferrite is shown in Figure 7.11.

In magnetite, Fe_3O_4, the divalent ions are iron and the interaction between ions on adjacent octahedral sites is particularly strong. One way of looking at the electronic structure of this oxide is to consider it as an array of O^{2-} ions and Fe^{3+} ions with the electrons that would have made half the Fe ions divalent delocalized over all the ions on octahedral sites. Fe^{3+} ions have five $3d$ electrons all with parallel spins. Since there can only be five $3d$ electrons of one spin on any atom, the delocalized spin

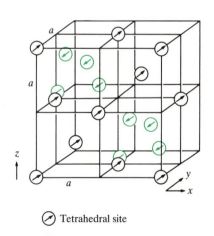

⊘ Tetrahedral site

⊘ Octahedral site

Figure 7.11 Magnetic structure of ferrimagnetic inverse spinel.

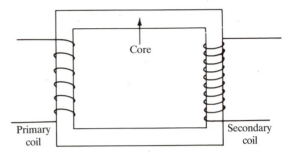

Figure 7.12 Sketch of the basic parts of a transformer.

must have the opposite spin. Being delocalized, it must also have the opposite spin to the $3d$ electrons on the next Fe ion. Hence the two ions must have their spins aligned, and these spins must be aligned with those of all other Fe ions on octahedral sites. Delocalization will be less for other ferrites.

Magnetite is the ancient lodestone used as an early compass. More recently ferrites have found use as memory devices in computers, as magnetic particles on recording tapes and as transformer cores. A simple sketch of the main parts of a transformer is shown in Figure 7.12. The importance of ferrites in transformer cores stems from their combination of low electrical conductivity, low coercive force and high magnetization. The low electrical conductivity, which is further reduced by laminating the core, reduces energy losses due to eddy currents. The high maximum magnetization ensures that the cores are not easily saturated and the low coercive force provides good linkage between the primary and secondary circuits.

7.7.1 Computer memory devices

In early computers, ferrites were employed as memory devices. For this purpose, they needed to switch cleanly between two different states which are read as 0 and 1 and to stay in one state until switched to the other. An almost square hysteresis curve (Figure 7.13) possessed by some ferrites, fulfils these requirements.

More recently, magnetic bubble devices have been used. These consist of very thin films of magnetic material which are manufactured to be magnetically anisotropic, i.e. their magnetization is not the same in different directions. In the absence of an applied field, magnetic domains magnetized up and down with respect to an axis at right angles to the film are formed (Figure 7.14). When a magnetic field in, say, the down direction is applied, the up domains will shrink and the down domains grow. As the applied field is increased, the up domains shrink further

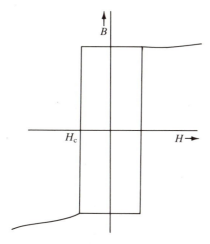

Figure 7.13 Idealized hysteresis curve of ferrite used for information storage.

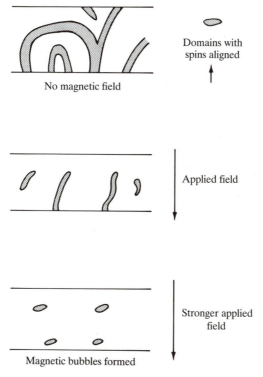

Figure 7.14 Domains in thin films giving rise to magnetic bubbles.

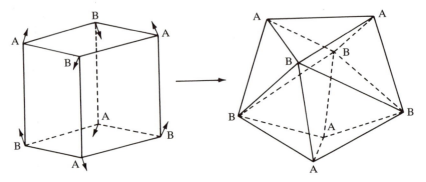

Figure 7.15 An eight-coordinate dodecahedral site.

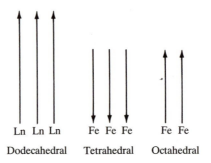

Figure 7.16 Alignment of spins in lanthanide iron garnet.

until they are only a few micrometres in diameter and become cylindrical.

These cylindrical domains are known as **magnetic bubbles** because their behaviour is described by equations similar to those used to describe soap bubbles. The films used are only thick enough to contain one bubble at any point. The bubbles move through the film towards regions of lower applied field and this effect is used to transmit information. A magnetic bubble is read as 1 and the absence of one as 0, giving a binary code as in other similar devices. Magnetic bubble devices can carry more information per unit area than older devices, but are being challenged by the optical storage devices as described in Chapter 6. The materials usually used are based on the garnet structure; a complex oxide structure with iron ions in octahedral and tetrahedral sites and lanthanide ions in (eight-coordinate) dodecahedral sites. Such an eight-coordinate site can be thought of as a distorted cube (Figure 7.15). The iron ions on the two types of site have their spins aligned antiparallel, and the lanthanide ions are aligned parallel to the iron ions on octahedral sites (Figure 7.16).

The anisotropy of the magnetism and the size of the bubbles is controlled by careful substitution of different ions, for example a garnet of composition $Sm_{0.4}Y_{2.6}Ga_{1.2}Fe_{3.8}O_{12}$ was used to develop 8 μm diameter bubble devices.

FURTHER READING

Cox, P.A. (1987) *Electronic Structure and Chemistry of Solids*, Chapter 5, Oxford University Press, Oxford.
See comments after Chapter 2. There is an extensive section on transition metal compounds.
West, A.R. (1988) *Basic Solid State Chemistry*, Chapter 8, John Wiley, New York.
See comments following Chapter 2.
Moore, W.J. (1967) *Seven Solid States*, Chapters IV and V, W.A. Benjamin Inc.
See comments following Chapter 2 (Chapter IV deals with ferromagnetism and Chapter V with antiferromagnetism).
Duffy, J.A. (1990) *Bonding, Energy Levels and Bands in Inorganic Solids*, Chapter 9, Longman, London.
See comments after Chapter 2. Concentrates on paramagnetism in metal complexes but also covers magnetic ordering.
Rao, C.N.R. and Gopalakrishnan, J. (1986) *New Directions in Solid State Chemistry*, Chapter 6, Cambridge University Press, Cambridge.
An interesting, though rather higher level, description of the interplay of structure and properties and their effect on applications.
Blunt, R. (1983) Magnetic bubble memories, *Chemistry in Britain*, September, 740.
A readable article explaining how these devices work. This issue of Chemistry in Britain *contains several articles relevant to solid state chemistry.*

QUESTIONS

1. Although manganese is not ferromagnetic, certain alloys such as Cu_2MnAl are ferromagnetic. The Mn–Mn distance in these alloys is greater than in manganese metal. What effect would this have on the 3d band of manganese? Why would this cause the alloy to be ferromagnetic?

2. The compound EuO has the NaCl structure and is paramagnetic above 70 K but magnetically ordered below it. Its neutron diffraction patterns at high and low temperatures are identical. What is the nature of the magnetic ordering?

3. $ZnFe_2O_4$ has the inverse spinel structure at low temperatures. What type of magnetism would you expect it to exhibit?

4. In transition metal pyrite disulphides, MS_2, the M^{2+} ions occupy octahedral sites. If a d band is formed, it will split into two as in the monoxides. Consider the information on some sulphides given below and decide whether the $3d$ electrons are localized or delocalized, which band the electrons are in if delocalized, and, in the case of semiconductors, between which two bands the band gap of interest lies.

MnS_2 antiferromagnetic (T_N = 78 K), insulator; above T_N paramagnetism fits five unpaired electrons per manganese.
FeS_2 diamagnetic, semiconductor.
CoS_2 ferromagnetic (T_C = 115 K), metal.

ANSWERS

1. Because the manganese atoms are further apart, the overlap of the $3d$ orbitals will be less. The $3d$ band will therefore be narrower than in manganese metal. With a narrower band, there is a larger inter-electronic repulsion and a state with a number of unpaired spins comparable to the number of atoms becomes favourable. The alloy is thus ferromagnetic.

2. The magnetically ordered unit cell is identical to the high-temperature (paramagnetic) unit cell, so that all the layers of europium ions must be aligned with their spins parallel, giving a ferromagnetic compound.

3. The Zn^{2+} and half the Fe^{3+} ions are on octahedral sites with spins aligned, and the remaining Fe^{3+} ions are on tetrahedral sites aligned antiparallel. The net moment of the Fe^{3+} ions is zero. As all the electron spins are paired in Zn^{2+} ions, there is no overall magnetic moment and the compound is antiferromagnetic.

4. In MnS_2 the $3d$ electrons are localized on the evidence given. Above the Néel temperature, the solid is an insulator with a paramagnetic susceptibility corresponding to five unpaired electrons per Mn and this could be explained on the basis of filled bands below the $3d$ level and localized $3d$ electrons. Below the Néel temperature, the localized electrons on different Mn ions interact to cancel out magnetic moments, possibly through a superexchange mechanism involving the disulphide ions.

    The properties of FeS_2 suggest that all spins are paired either in a band or localized on the iron ions. There are six $3d$ electrons per Fe ion and this is just enough to fill the lower t_{2g} band. The semiconducting properties suggest that there is an empty band not much higher in energy. This is probably the e_g band.

In CoS_2, there are enough electrons to partly occupy the e_g band and thus this compound is metallic. The band is narrow and so gives rise to ferromagnetism.

<table>
<tr><td>8</td><td></td></tr>
</table>

8 Superconductivity

8.1 INTRODUCTION

In the late 1980s the amount of research effort in the field of superconductors increased dramatically due to the discovery of so-called 'high-temperature' superconductors by Bednorz and Muller in 1986. Their findings were thought to be so important that it was only a year later that they were awarded the Nobel Prize for Physics by the Royal Swedish Academy of Sciences.

Superconductors have two unique properties that could be very important for technology if methods can be found for exploiting them. First, they have zero electrical resistance and so carry current with no energy loss: this could revolutionize the national grid for instance, and is already exploited in the windings of superconducting magnets as used in NMR experiments. Second, they expel all magnetic flux from their interior and so are forced out of a magnetic field. The superconductors can float or 'levitate' above a magnetic field: the Japanese have an experimental frictionless train that floats above magnetic rails and has achieved speeds of over $500\,\mathrm{km\,h^{-1}}$ (300 mph). The possibilities are exciting, but until recently there has always been the major snag that superconductivity was observed only at temperatures close to absolute zero. Until 1986, the highest temperature a superconductor operated at was 23 K, and so they all had to be cooled by liquid helium (boiling temperature $\sim 4\,\mathrm{K}$). This, of course, made any use of superconductors extremely expensive.

For many years workers have sought to find a superconductor that will operate at the very least at liquid nitrogen temperatures (above 77 K) so as to be commercially viable, hoping, eventually for a room-temperature superconductor. The events of recent years have brought the operating temperatures above the 'magic' liquid nitrogen temperature, and there may be evidence for superconductivity being observed above 200 K. No one has yet reported a room-temperature superconductor.

8.2 THE DISCOVERY OF SUPERCONDUCTORS

In 1908 Kamerlingh Onnes succeeded in liquefying helium, and this paved the way for many new experiments to be performed on the behaviour of materials at low temperatures. For a long time, it had been known from conductivity experiments that the electrical resistance of a metal decreased with temperature. In 1911, Onnes was measuring the variation of the electrical resistance of mercury with temperature when he was amazed to find that at 4.2 K the resistance suddenly dropped to zero. He called this effect **superconductivity** and the temperature at which it occurs is known as the (**superconducting**) **critical temperature**, T_c. This effect is illustrated for tin in Figure 8.1. One effect of the zero resistance is that there is no power loss in an electrical circuit made from a superconductor. Once an electrical current is established, it shows no discernible decay for as long as experimenters have been able to watch – a maximum of $2\frac{1}{2}$ years so far!

For more than 20 years, little progress was made in the understanding of superconductors and only more substances exhibiting the effect were found. More than 20 metallic elements can be made superconducting under suitable conditions (Figure 8.2), as can thousands of alloys. It was not until 1933 that a new effect was observed by Meissner.

8.3 THE MAGNETIC PROPERTIES OF SUPERCONDUCTORS

Meissner and Ochsenfeld found that when a superconducting material is cooled below its critical temperature, T_c, it expels all magnetic flux from within its interior (Figure 8.3a); the magnetic flux, B, is thus zero inside a superconductor. Since $B = \mu_0 H(1 + \chi)$, when $B = 0$, χ must equal -1,

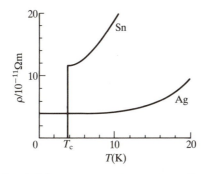

Figure 8.1 A plot of resistivity, ρ versus temperature, T, showing the drop to zero at the critical temperature, T_c, for a superconductor, and the finite resistivity of a normal metal at absolute zero.

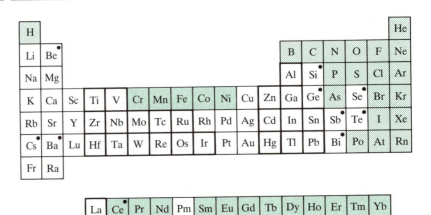

Figure 8.2 The superconducting elements.

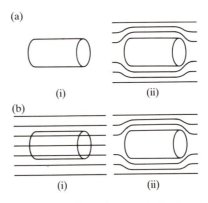

Figure 8.3 (a) (i) Superconductor with no magnetic field. When a field is applied in (ii) the magnetic flux is excluded. (b) (i) Superconducting substance above the critical temperature, T_c, in a magnetic field. When the temperature drops below the critical temperature (ii), the magnetic flux is expelled from the interior. Both are called Meissner effects.

Figure 8.4 A permanent magnet floating over a superconducting surface.

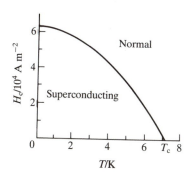

Figure 8.5 The variation with temperature of the critical field strength, H_c, for lead. Note that H_c is zero when the temperature, T, equals the critical temperature, T_c.

i.e. superconductors are perfect diamagnets. If a magnetic field is applied to a superconductor, the magnetic flux is excluded (Figure 8.3b) and the superconductor repels a magnet. This is shown in Figure 8.4, where a magnet is seen floating in mid-air above a superconductor.

It is also found that T_c, changes in the presence of a magnetic field. A typical plot of T_c against increasing magnetic field is shown in Figure 8.5; as the applied field increases, the critical temperature drops. It follows that a superconducting material can be made non-superconducting by the application of a large enough magnetic field. The minimum value of the field strength required to bring about this change is called the **critical field strength, H_c**; its value depending on the material in question and on the temperature.

Similarly, if the current in the superconductor exceeds a **critical current**, the superconductivity is destroyed. This is known as the **Silsbee effect**. The size of the critical current is dependent on the nature and geometry of the particular sample.

8.4 THE THEORY OF SUPERCONDUCTIVITY

The theory of superconductivity is extremely complex, lying firmly in the realm of theoretical physics. This section attempts to give a qualitative picture of the ideas involved and also to give some familiarity with the terminology.

Physicists worked for many years to find a theory that explained superconductivity. To begin with, it looked as though the lattice played no part in the superconducting mechanism as X-ray studies showed that there was no change in either the symmetry or the spacing of the lattice when superconductivity occurred. However, in 1950, an **isotope effect** was first observed: for a particular metal, the critical temperature was found to depend on the isotopic mass, M, such that:

$$T_c \; \alpha \; \frac{1}{\sqrt{M}}.$$
(8.1)

The frequency, ν, of vibration of a diatomic molecule is known to be given by:

$$\nu = \frac{1}{2\pi} \sqrt{\frac{k}{\mu}},$$
(8.2)

where μ is the reduced mass of the molecule, and k the force constant of the bond. We can see that a vibration also changes frequency on isotopic substitution such that the frequency, ν, is proportional to $1/\sqrt{\text{mass}}$. This suggested to physicists that superconductivity was in some way related to the vibrational modes of the lattice and not just to the conduction electrons. The vibrational modes of a lattice are quantized, as are the modes of an isolated molecule, the quanta of the lattice vibrations being called **phonons**.

Frohlich suggested that there could be a strong phonon–electron interaction in a superconductor that leads to an attractive force between two electrons that is strong enough to overcome the Coulomb repulsion between them. Very simply, the mechanism works like this: as a conduction electron passes through the lattice, it can disturb some of the positively charged ions from their equilibrium positions, pushing them together and giving a region of increased positive charge density. As these oscillate back and forth, a second electron passing this moving region of increased positive charge density is attracted to it. The net effect is that the two electrons have interacted with one another, using the lattice vibration as an intermediary. Furthermore, the interaction between the electrons is attractive because each of the two separate steps involved an attractive Coulomb interaction.

It is the scattering of conduction electrons by the lattice vibrations, phonons, that produces electrical resistance at room temperature. (At low temperatures it is predominantly the scattering by lattice defects that gives electrical resistance.) Contrary to what we might have expected intuitively, a superconductor will have high resistance at room temperature, because it has strong electron–phonon interactions. Indeed, the best room-temperature electronic conductors, silver and copper, do not superconduct at all! Superconductors do not have low electrical resistance above the superconducting T_c.

In 1957, Bardeen, Cooper and Schrieffer published their theory of superconductivity, known as the **BCS theory**. It predicts that under certain conditions, the attraction between two conduction electrons due to a succession of phonon interactions can slightly exceed the repulsion that they exert directly on one another due to the Coulomb interaction of their like charges. The two electrons are thus weakly bound together forming a so-called **Cooper pair**. It is these Cooper pairs that are responsible for superconductivity.

BCS theory shows that there are several conditions that have to be met for a sufficient number of Cooper pairs to be formed and superconductivity to be achieved. It is beyond the scope of this book to go into this in any depth. Suffice it to say that the electron–phonon interaction must be strong and that low temperature favours pair formation, hence high-temperature superconductors are not predicted by BCS theory.

Cooper pairs are weakly bound, with typical separations of 10^6 pm for the two electrons. They are also constantly breaking up and reforming (usually with other partners). There is thus enormous overlap between different pairs and the pairing is a complicated dynamic process. The ground state of a superconductor therefore is a 'collective' state, describing the ordered motion of large numbers of Cooper pairs. When an external electrical field is applied, the Cooper pairs move through the lattice under its influence. However, they do so in such a way that the ordering of the pairs is maintained. The motion of each pair is locked to the motion of all the others, and none of them can be individually scattered by the lattice. Because the pairs cannot be scattered by the lattice the resistance is zero and the system is a superconductor.

8.5 JOSEPHSON EFFECTS

In 1962, Josephson predicted that if two superconducting metals were placed next to each other separated only by a thin insulating layer (such as their surface oxide coating) then a current would flow in the absence of any applied voltage. This effect is indeed observed because if the barrier is not too thick then electron pairs can cross the junction from one superconductor to the other without dissociating. This is known as the **d.c. Josephson effect**. He further predicted that the application of a small d.c. electric potential to such a junction would produce a small alternating current, the **a.c. Josephson effect**. These two properties are of great interest to the electronics and computing industries where they can be exploited for fast-switching purposes.

8.6 THE SEARCH FOR A HIGH-TEMPERATURE SUPERCONDUCTOR

By 1973, the highest temperature found for the onset of superconductivity was 23.3 K. This was for a compound of niobium and germanium, Nb_3Ge, and here it stayed until 1986 when Georg Bednorz and Alex Müller reported their findings. Their Nobel Prize citation states 'Last year, 1986, Bednorz and Müller reported finding superconductivity in an oxide material at a temperature 12°C higher than previously known.' The compound that prompted their initial paper has been shown to be $La_{2-x}Ba_xCuO_4$, where $x = 0.2$, with a structure based on that of K_2NiF_4, a perovskite-related layer compound: they observed the onset of superconductivity at 35 K. (We will look at the detailed structure in the next section.) The insight that they brought to this field was to move away from the investigation of metals and their alloys and to study systematically the solid state physics and chemistry of metallic oxides.

The idea was soon born that it might be possible to raise the temperature even further by substitution with different metals. Using this technique, it was Chu's group at Houston, Texas that finally broke through the liquid nitrogen temperature barrier with the superconductor that is now known as '1–2–3'. This superconductor replaces lanthanum with yttrium and has the formula $YBa_2Cu_3O_{7-x}$. The onset of superconductivity for 1–2–3 occurs at 93 K.

8.7 THE CRYSTAL STRUCTURES OF HIGH-TEMPERATURE SUPERCONDUCTORS

The perovskite structure is named after the mineral $CaTiO_3$ (Chapter 1, section 1.5.3). Many oxides of general formula ABO_3 adopt this structure

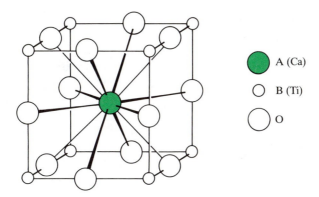

Figure 8.6 The A-type unit cell of the perovskite structure for compounds ABO_3, such as $CaTiO_3$.

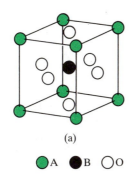

(a)

A   B   O

Figure 8.7 B-type cell for the ABO_3 perovskite structure.

(also fluorides, ABF_3 and sulphides, ABS_3). The so-called perovskite A-type unit cell (with the A-type atom in the centre of the cell) is shown in Figure 8.6. The central A atom (Ca) is coordinated by eight titaniums at the corners of the cube and by 12 oxygens at the mid-points of the edges. The perovskite structure can be equally well represented by moving the origin of the unit cell to the body centre. This has the effect of putting Ca (A) atoms at each corner, Ti (B) atoms at the body centre and an O atom in the centre of each face (Figure 8.7).

The K_2NiF_4 structure is adopted by the Bednorz/Müller superconductor $La_{2-x}Ba_xCuO_4$. Its unit cell, and its relationship to the perovskite structure is illustrated in Figure 8.8. The structure has a body-centred tetragonal cell. We can get more insight into the structure by dividing it up into three sections along its c direction (Figure 8.8a). The central

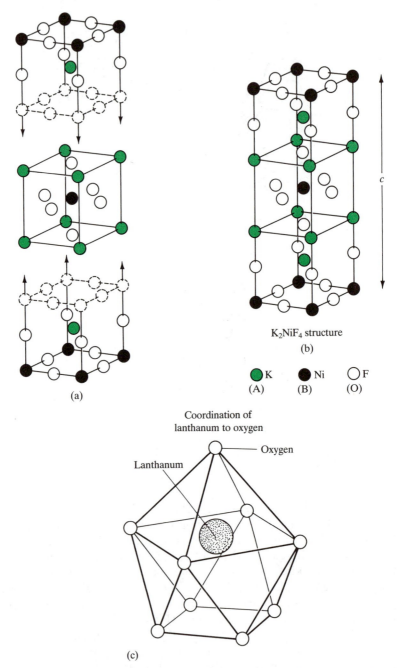

Figure 8.8 (a) and (b) show the K_2NiF_4 structure derived from perovskite-type (ABO_3) unit cells, as adopted by $La_{2-x}Ba_xCuO_4$. (c) The nine-fold coordination of lanthanum by oxygen in this structure.

section is a B-type perovskite unit cell, and this is sandwiched by two A-type cells that have their bottom and top layers, respectively, missing (they are shown by dotted atomic positions). The lanthanum and barium atoms are similar in size and they are randomly distributed amongst the K (A) atom positions in the structure. Cu takes the Ni (B) positions. Notice that the A-type atoms (K or La/Ba) are surrounded only by nine oxygens in this structure whereas A was coordinated by 12 equidistant oxygens in the basic perovskite structure.

For the sake of simplicity each section of the unit cell has been shown as a cube. In practice the central section is slightly elongated in the c direction and the Cu atoms are coordinated by an elongated octahedron of six Os (and by eight La/Bas at the corners). This elongation is a Jahn–Teller effect and is the reason that Bednorz and Müller chose to investigate these copper oxides for superconductivity. The copper atoms and the four closest oxygen atoms lie in the ab planes forming networks of copper and oxygen atoms separated by planes of other atoms. It is this planar arrangement of copper and oxygen atoms that confers the super-conducting properties on the crystal.

The parent compound, La_2CuO_4 (without any barium in the lattice) has no superconductivity. This is because the parent compound is anti-ferromagnetic. This crystal contains Cu^{2+} ions and the unpaired electrons on the copper ions align antiparallel throughout the structure. The presence of this interaction locks the electrons to the lattice stopping both conductivity and superconductivity very effectively. When some barium (Ba^{2+}) is introduced into the lattice replacing lanthanum (La^{3+}) obviously some charge compensation has to be made in order to preserve electrical neutrality. For every Ba^{2+} present one Cu^{II} must be oxidized to Cu^{III} and this breaks up the alignment of the unpaired spins. When the average valence of the copper atoms reaches a critical value of ~2.2 the antiferromagnetism disappears and superconductivity appears. (Substitution with strontium or calcium also produces superconductivity.) It has been observed that if pressure is applied to the compound, such that the Cu–O distance decreases, then T_c increases. There is also a general correlation of higher T_c with higher Cu(III) content: a Cu(III)–O bond is shorter than a Cu(II)–O bond because the electron is removed from an antibonding e_g orbital. These two observations are not necess-arily related because the high-pressure studies show an increase of T_c with decrease in Cu–O distance, without any change in the Cu(III) content of the compound.

The crystal structure of the 1–2–3 superconductor, $YBa_2Cu_3O_{7-x}$, is shown in Figure 8.9. Figure 8.9a shows only the positions of the metal atoms: note the strong similarity between this and the structure of $La_{2-x}Ba_xCuO_4$. If, as before, we discuss it in terms of the perovskite structure ABO_3, where B = Cu, the central section is now an A-type

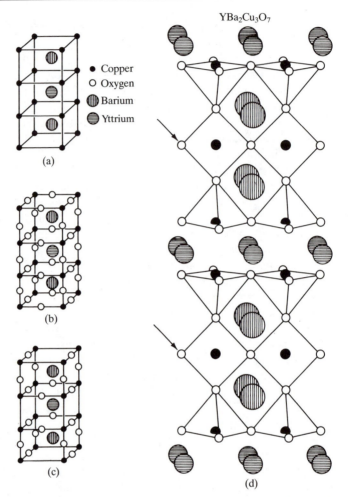

YBa$_2$Cu$_3$O$_7$

● Copper
○ Oxygen
◫ Barium
◒ Yttrium

(a)

(b)

(c)

(d)

Figure 8.9 The structure of 1–2–3: (a) the metal positions; (b) an idealized unit cell of the hypothetical YBa$_2$Cu$_3$O$_9$, based on three perovskite A-type unit cells; (c) idealized structure of YBa$_2$Cu$_3$O$_{7-x}$; (d) the structure of YBa$_2$Cu$_3$O$_7$, showing copper–oxygen planes formed by the bases of the pyramids, with the copper–oxygen diamonds in between.

perovskite unit cell and above and below it are also A-type perovskite unit cells with their bottom and top layers missing. This gives copper atoms at the unit cell corners and on the unit cell edges at fractional coordinates $\frac{1}{3}$ and $\frac{2}{3}$. The atom at the body centre of the cell (i.e. in the centre of the middle section) is now **yttrium**. The atoms in the centres of the top and bottom cubes are **barium**.

If, in this structure, all three sections were based exactly on perovskite unit cells, we would expect to find the oxygen atoms in the middle of each cube edge (Figure 8.9b). This would give an overall formula of $YBa_2Cu_3O_9$. This formula is improbable because it gives an average oxidation state for the three copper atoms of $\frac{11}{3}$ implying that the unit cell contains both Cu(III) and Cu(IV), which is unlikely as Cu(IV) complexes are extremely rare. The unit cell in fact contains only approximately seven oxygen atoms ($YBa_2Cu_3O_{7-x}$). When $x = 0$, the oxygen atoms on the vertical edges of the central cube are not there and there are also two missing from both the top and bottom faces (Figure 8.9c). A unit cell containing seven oxygen atoms has an average copper oxidation state of 2.33 indicating the presence of Cu(II) and Cu(III) in the unit cell (but no longer Cu(IV)).

In the 1–2–3 structure (when $x = 0$) the yttrium atom is coordinated by eight oxygens and the barium atoms by 10 oxygens. The oxygen vacancies in the 1–2–3 superconductor create sheets and chains of linked copper and oxygen atoms running through the structure (this is shown slightly idealized in the diagrams as, in practice, the copper atoms lie slightly out of the plane of the oxygens); the copper is in four-fold square-planar or five-fold square-pyramidal coordination (Figure 8.9d). The superconductivity is found in directions parallel to the copper planes which are created by the bases of the Cu–O pyramids and which are separated by layers of yttrium atoms. These Cu–O nets seem to be a common feature of all the new high-temperature superconductors. If this superconductor is made more deficient in oxygen, at $YBa_2Cu_3O_{6.5}$ ($x = 0.5$) the superconducting T_c, drops to 60 K and at $YBa_2Cu_3O_6$ the superconductivity disappears. The oxygen is not lost at random but goes from specific sites, gradually changing the square-planar coordination of the copper along the c direction into the two-fold linear coordination characteristic of Cu^+. The sites from which the oxygen atoms are lost are indicated by arrows in Figure 8.9d. The arrangement of the copper and oxygen atoms in the base of the pyramids is not affected. However, when the formula is $YBa_2Cu_3O_6$ all the square planar units along c have become chains containing Cu^+ and the pyramid bases contain only Cu^{2+}; the unpaired spins of the Cu^{2+} are aligned antiparallel and the compound is antiferromagnetic. It is not until the oxygen content is increased to $YBa_2Cu_3O_{6.5}$ that the antiferromagnetic properties are destroyed and the compound becomes a superconductor. It is thought that this compound contains copper in all three oxidation states: I, II and III. $YBa_2Cu_3O_7$ contains Cu(II) and Cu(III) both in the sheets and in the chains. Clearly the oxidation state of the coppers in the structure (and thus their bonding connections and bond lengths) is extremely important in determining both, whether superconductivity occurs at all, and the temperature below which it occurs (T_c).

In both of the superconductors discussed here the average oxidation state of the copper is greater than two and as a result positive holes are formed in the valence bands. The charge is carried by the positive holes and as a consequence these materials are known as **p-type supercon-ductors**. Until 1988 all the high-temperature superconductors that had been found were p-type and it was assumed by many that this would be a feature of high-temperature superconductors. However, some **n-type superconductors** have now also been discovered, where the charge carriers are electrons. The first to be found was based on the compound Nd_2CuO_4 with small amounts of the three-valent neodymium substituted by four-valent cerium, $Nd_{2-x}Ce_xCuO_{4-y}$ where $x \sim 0.17$ (samarium, europium or praseodymium can also be substituted for the neodymium). Other similar compounds have since been found based on this structure where the three-valent lanthanide is substituted by, for example, four-valent thorium in $Nd_{2-x}Th_xCuO_{4-y}$. The superconductivity occurs at $T_c \leqslant 25\,K$ for these compounds.

At the time of writing (1991) the superconductors with the highest known critical temperatures are formed from thallium, barium, calcium, bismuth, copper and oxygen. Again the characteristic Cu–O planes are present, interspersed by ones containing thallium and oxygen. The highest T_c reported so far for this series of compounds is $125\,K$ for $Tl_2Ba_2CaCu_3O_{10}$.

It seems clear that these new ceramic superconductors all have a common feature: the presence of Cu–O layers sandwiched between layers of other elements. The superconductivity takes place in these planes and the other elements present and the spacings between the planes changes the superconducting transition temperature; exactly how is not yet understood. Theoretical physicists have been working hard to try and explain the phenomenon of high-temperature superconductivity, but with no clear consensus so far. It has been hoped that the later discovery of n-type superconductors will shed some light on the problem. BCS theory does not account for all the evidence concerning high-temperature superconductors: first, the net attraction resulting from the lattice does not appear to be strong enough to account for the high T_c values and second, although there is evidence that pairs of electrons are still responsible for high-temperature superconductivity, it is not yet clear whether they interact with each other via a lattice vibration, because the isotopic substitution data is contradictory.

8.8 POTENTIAL USES OF HIGH-TEMPERATURE SUPERCONDUCTORS

In order to be useful in many of the possible applications of super-conductors, the new high-temperature superconductors have to meet

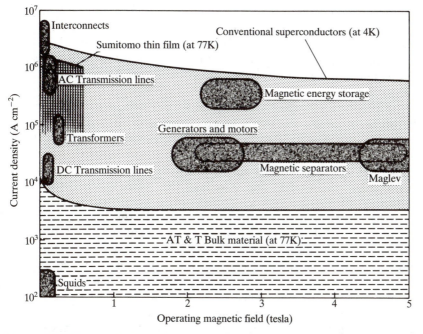

Figure 8.10 The graph shows the necessary performance of superconductors for various devices, based on current designs. Underneath is drawn the capability of both high- and low-temperature superconductors that can currently (1991) be manufactured.

some fairly stringent criteria: they have to be able to carry currents of more than $10^9 \, \mathrm{A \, m^{-2}}$ ($100\,000 \, \mathrm{A \, cm^{-2}}$); withstand strong magnetic fields (up to $5 \, \mathrm{T}$); and be physically strong enough to be used in motors and generators. Some of the criteria for these different uses are summarized in Figure 8.10. For some of the uses, bulk samples will not be useful and the material will need to be available in the form of wire, tape or thin film. Added to this, the benefits of using the new materials (better performance or cheaper running costs, for instance) have to outweigh any extra capital costs that may be needed to change to the new technology.

Making useful products in the form of wire or tape has proved to be difficult. They are made from a powdered mixture of superconductor and polymer which is formed into the appropriate shape, sintered, then annealed; unfortunately the product tends to be brittle. Another problem is that the 1–2–3 superconductor tends to lose oxygen, and therefore its superconductivity at the same time; other ceramics may not have this same problem. It has proved possible to make thin films that can carry large currents – of the order of $10^6 \, \mathrm{A \, cm^{-2}}$ ($10^{10} \, \mathrm{A \, m^{-2}}$) – and it may be feasible to mount these films on a backing tape that can be wound or shaped.

The uses of superconductors are based around four main properties: the Meissner effect, the lack of electrical resistance, the ability to create superconducting magnets and the Josephson effect. We will look at each area in turn.

The property of superconductors that has received most publicity is the Meissner effect, the ability to expel magnetic flux from the interior and so float or levitate above a magnet. The Japanese have been developing a magnetically levitating train, the so-called '**maglev**', using low-temperature superconducting magnets. Use of the new high-temperature superconductors instead, should reduce the refrigeration costs of the project. Unfortunately refrigeration is only a very small percentage of both the capital and running costs and so will probably not make the difference as to whether this becomes a viable mode of transport or not. The hope is to produce a train that can travel at several hundreds of kilometres per hour and so compete with flying over short and medium length journeys.

Conventional electromagnets (made by passing an electric current through insulated copper wire wound around an iron core) can produce fields of about two tesla. Low-temperature superconducting magnets are manufactured from both NbTi and Nb_3Sn wires which can carry much higher currents than copper wire ($\sim 400\,000\,A\,cm^{-2}$) and so eliminate the need for the amplifying iron core. Such superconducting magnets are both lighter and more powerful ($\sim 10\,T$) than their conventional counterparts, consume very little power, and produce little heat, but they have to be cooled to 4 K to become superconducting. If the windings could be replaced by high-temperature superconductors the liquid helium coolant (priced in £s per litre) could be replaced by liquid nitrogen (priced in pence per litre) thus dramatically reducing the running costs. It might also be possible to produce even larger value magnetic fields than currently achieved.

Low-temperature superconducting magnets are currently employed for several important uses. Strong magnets are used to remove impurities from food and raw materials; for instance, the magnetic impurities in china clay discolour the manufactured product if not removed. The latest NMR spectrometers employ superconducting magnets to provide large magnetic fields, so improving sensitivity. They are an important tool in fundamental scientific research, and are increasingly being used in body scanners for NMR imaging used in medical diagnosis. However, the prospect of high-temperature supermagnets still seems a long way off, until a wire is developed that can carry the kind of current densities possible for the niobium alloys.

Electrical generators produce power by spinning a magnet inside a coil;

an electrical current is produced in the coil. It is thought that the use of a superconducting magnet would make the process more efficient, and the electricity a little cheaper. As with many of the other possible uses for the high-temperature superconductors, the percentage savings would be quite small, perhaps 1%, and whether this will ever offset the immense capital investment required is not clear.

Although superconductors carry direct current (d.c.) with no loss of power and without generating heat, there is a small loss of power when they carry alternating current (a.c.) due to the production of radio waves. This absence of electrical resistance could be extremely useful in the distribution of electricity throughout the country, as the copper and aluminium wires now used lose 5–8% of the power due to the resistance of these elements. The superconducting wire (once it could be manufactured) is unlikely to be strong enough to be suspended from pylons, this means that all the cables, together with the cooling system, would have to be at ground level or buried underground – an expensive capital outlay! On the other hand, the environmental pressure groups would be greatly in favour of this, both for aesthetic and for safety reasons.

A supermagnetic energy storage unit (**SMES**) has seen designed that could store a massive amount of direct current with no energy loss. The design currently uses NbTi alloy and a liquid helium cooling system. It would be used to store electricity generated at times of low demand, at night or in the summer for instance. A liquid nitrogen cooling system would be cheaper to build and to run.

Josephson junctions have been of great interest to the electronics industry because they can be used for switching voltages very quickly. As they do so, they consume very little power and thus do not heat up. This means that they can be packed very close together without the need for elaborate cooling, giving the potential for making smaller and faster computers. IBM expended much time and research on building a Josephson computer based on these junctions. Sadly the project was eventually abandoned because they found the memory cells unreliable. Japanese workers are still continuing the research, but in the meantime semiconductor technology reigns supreme.

Josephson junctions are used in Superconducting Quantum Interference Devices (**SQUIDs**). These consist of a loop of superconductive wire with either one built-in Josephson junction (RF SQUID) or two (DC SQUID). The device is extremely sensitive to changes in magnetic field, and can measure voltages as small as 10^{-18} V, currents of 10^{-18} A and magnetic fields of 10^{-14} T. SQUIDs are being used in medical research to detect small changes in magnetic field in the brain. Geologists are able to employ SQUIDs in prospecting for minerals and oil where deposits can cause small local changes in the earth's magnetic field; they also use the device for research into plate tectonics, collecting magnetic data on rocks of different ages. Physicists use SQUIDs for research into fundamental particles, and it is thought that there may be military applications in detecting the change in the earth's magnetic field caused by submarines. Clearly SQUIDs with their remarkable sensitivity have a myriad of potential uses. A change to high-temperature superconductors in the devices could make them cheaper and more flexible to operate.

Superconductivity has been a very active research area. Japanese workers are pouring huge resources into it, and there are many active research groups in other continents both in industry and in the universities. The rewards are potentially great if the high-temperature superconductors can be made economic. The rewards could be even greater if a room-temperature superconductor can be found. As we have seen, there have already been four Nobel prizes awarded to people working in this area: to Kamerlingh Onnes for the initial discovery; to Bardeen, Cooper and Schrieffer for BCS theory; to Josephson; and most recently to Bednorz and Müller. The work so far has been very exciting, we can only hope that in the future, superconductivity, together with the rest of the solid state field, holds even more wonderful discoveries for us.

FURTHER READING

This field has moved very rapidly in recent years and this chapter may well date quite rapidly. To keep up with current developments the reader is referred to journals such as **Angewandte Chemie, New Scientist, Science** *and* **Scientific American**.

Vanderah, T. (ed.) *Chemistry of Superconductor Materials*, Noyes Publications (in press).

Ellis, A.B. (1987) Superconductors: better levitation through chemistry, *J. Chem. Ed.*, **64**, 836–41.

Cava, R.J. (1990) Superconductors beyond 1–2–3, *Scientific American*, August, 24–31.

Hazen, R.M. (1990) Perovskites, *Scientific American*, June, 52–61.

Grant, P. (1987) Do-it-yourself superconductors, *New Scientist*, 30 July, 36–39.

Lamb, J. (1987) Industry warms to superconductivity, *New Scientist*, 22 October, 56–61.

Wolsky, A.M., Giese, R.F. and Daniels, E.J. (1989) The new superconductors: prospects for applications, *Scientific American*, February, 45–52.

Cava, R.J. (1990) Structural chemistry and the local charge picture of copper oxide superconductors, *Science*, **247**, 656–62.

Sleight, A.W. (1988) Chemistry of high-temperature superconductors, *Science*, **242**, 1519–27.

Khurana, A. (1989) Electron superconductors challenge theories, start a new race, *Physics Today*, April, 17–19.

QUESTIONS

1. A party of late-night revellers, returning home, decide to take a short cut across a field. Unfortunately, the night is dark and moonless and the field is known to contain deep potholes. Someone has the bright idea that if everyone links arms and advances together across the field, then if one of the company does encounter a pothole she will be lifted clear by dint of collective support! Taking this story as an analogy (not a perfect one of course) with BCS theory, try to identify as many components as possible with corresponding components of BCS theory.

2. Draw a packing diagram (Chapter 1, section 1.4.5) of the perovskite A-type cell ($CaTiO_3/ABO_3$), determine the number of ABO_3 formula units, and describe the coordination geometry around each type of atom. Repeat this procedure for the perovskite B-type unit cell.

3. The unit cell of the K_2NiF_4 crystal structure is shown in Figure 8.8. Draw packing diagrams for layers at fractional coordinates $\frac{1}{6}$ and $\frac{1}{3}$ in c.

ANSWERS

1. In the analogy, each person represents a Cooper pair and the linking of arms denotes the overlap between the pairs, giving an ordered system. Because of the ordered collective motion, scattering from defects – tripping over potholes – cannot take place! The analogy breaks down in that it allows for overlap only between adjacent pairs rather than over large numbers; also, the loss of superconductivity leads to the scattering of individual electrons, not of individual Cooper pairs.

2. The packing diagram for the A-type unit cell is shown in Figure 8.11.

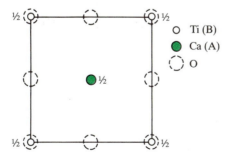

Figure 8.11 Packing diagram for a unit cell of perovskite, $ABO_3/CaTiO_3$.

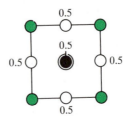

Figure 8.12 Packing diagram for a perovskite B-type cell.

The contents of the unit cell are as follows: 12 O atoms midway on cell edges, each shared by four unit cells; one Ca atom at the body centre, eight Ti atoms at the cube corners, each shared by eight unit cells : $[(12 \times \frac{1}{4})O + 1Ca + (8 \times \frac{1}{8})Ti] = 3O + 1Ca + 1Ti = 1$ formula unit. The Ca in the centre is surrounded by 12 O atoms, and eight Ti atoms at the corners of a cube. A Ti atom is surrounded octahedrally by six O atoms, and by eight Ca atoms at the corners of a cube. An O atom is linearly coordinated by two Ti atoms and by four Ca atoms in a square-planar configuration.

The packing diagram for the B-type cell is shown in Figure 8.12. This has the Ti/B atom at the centre of the cell and the Ca/A atoms at the corners, each shared by eight other unit cells; there are six face-centred oxygen atoms, each shared by two unit cells. Counting up as before gives: $[(8 \times \frac{1}{8})Ca + 1Ti + (6 \times \frac{1}{2})O] = 1Ca + 1Ti + 3O = 1$ formula unit. The coordination geometry of course remains identical; we have only moved the origin of the unit cell.

3. The packing diagrams for the layers at $\frac{1}{6}$ and $\frac{1}{3}$ (and $\frac{5}{6}$ and $\frac{2}{3}$) in c are shown in Figure 8.13.

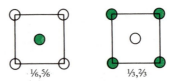

Figure 8.13 Layer sequences for K_2NiF_4.

Basic SI units

Physical quantity (and symbol)	Name of SI unit	Symbol for unit
Length (l)	metre	m
Mass (m)	kilogram	kg
Time (t)	second	s
Electric current (I)	ampere	A
Thermodynamic temperature (T)	kelvin	K
Amount of substance (n)	mole	mol
Luminous intensity (I_v)	candela	cd

Derived SI units having special names and symbols

Physical quantity (and symbol)	Name of SI unit	Symbol for SI derived unit and definition of unit
Frequency (v)	hertz	$Hz\ (= s^{-1})$
Energy (U), enthalpy (H)	joule	$J\ (= kg\,m^2\,s^{-2})$
Force	newton	$N\ (= kg\,m\,s^{-2} = J\,m^{-1})$
Power	watt	$W\ (= kg\,m^2\,s^{-3} = J\,s^{-1})$
Pressure (p)	pascal	$Pa\ (= kg\,m^{-1}\,s^{-2} = N.m^{-2} = J\,m^{-3})$
Electric charge (Q)	coulomb	$C\ (= A\,s)$
Electric potential difference (V)	volt	$V\ (= kg\,m^2\,s^{-3}\,A^{-1} = J\,A^{-1}\,s^{-1})$
Capacitance (c)	farad	$F\ (= A^2\,s^4\,kg^{-1}\,m^{-2} = A\,s\,V^{-1} = A^2\,s^2\,J^{-1})$
Resistance (R)	ohm	$\Omega\ (= V\,A^{-1})$
Conductance (G)	siemen	$S\ (= A\,V^{-1})$
Magnetic flux density (B)	tesla	$T\ (= V\,s\,m^{-2} = J\,C^{-1}\,s\,m^{-2})$

SI prefixes

10^{-15}	10^{-12}	10^{-9}	10^{-6}	10^{-3}	10^{-2}	10^{-1}	10^{3}	10^{6}	10^{9}	10^{12}
femto	pico	nano	micro	milli	centi	deci	kilo	mega	giga	tera
f	p	n	μ	m	c	d	k	M	G	T

Fundamental constants

Constant	Symbol	Value
Speed of light in a vacuum	c	$2.997925 \times 10^8 \, \mathrm{m\,s^{-1}}$
Charge of proton	e	$1.602189 \times 10^{-19} \, \mathrm{C}$
Charge of electron	$-e$	
Avogadro constant	N_A	$6.022045 \times 10^{23} \, \mathrm{mol^{-1}}$
Boltzmann constant	k	$1.380662 \times 10^{-23} \, \mathrm{J\,K^{-1}}$
Gas constant	$R = N_A k$	$8.31441 \, \mathrm{J\,K^{-1}\,mol^{-1}}$
Faraday constant	$F = N_A e$	$9.648456 \times 10^4 \, \mathrm{C\,mol^{-1}}$
Planck constant	h	$6.626176 \times 10^{-34} \, \mathrm{J\,s}$
	$\hbar = \dfrac{h}{2\pi}$	$1.05457 \times 10^{-34} \, \mathrm{J\,s}$
Vacuum permittivity	ε_0	$8.854 \times 10^{-12} \, \mathrm{F\,m^{-1}}$
Vacuum permeability	μ_0	$4\pi \times 10^{-7} \, \mathrm{J\,s^2\,C^{-2}\,m^{-1}}$
Bohr magneton	μ_B	$9.27402 \times 10^{-24} \, \mathrm{J\,T^{-1}}$
Electron g value	g_e	2.00232

Miscellaneous physical quantities

Name of physical quantity	Symbol	SI unit
Enthalpy	H	J
Entropy	S	$J\,K^{-1}$
Gibbs function	G	J
Standard change of molar enthalpy	$\Delta H_m^{\ominus}$	$J\,mol^{-1}$
Standard change of molar entropy	$\Delta S_m^{\ominus}$	$J\,K^{-1}\,mol^{-1}$
Standard change of molar Gibbs function	$\Delta G_m^{\ominus}$	$J\,mol^{-1}$
Wavenumber	$\sigma\left(=\dfrac{1}{\lambda}\right)$	cm^{-1}
Atomic number	Z	dimensionless
Conductivity	σ	$S\,m^{-1}$
Density	ρ	$kg\,m^{-3}$
Molar bond dissociation energy	D_m	$J\,mol^{-1}$
Molar mass	$M\left(=\dfrac{m}{n}\right)$	$kg\,mol^{-1}$

The Greek alphabet

alpha	A	α	nu	N	ν
beta	B	β	xi	Ξ	ξ
gamma	Γ	γ	omicron	O	o
delta	Δ	δ	pi	Π	π
epsilon	E	ε	rho	P	ρ
zeta	Z	ζ	sigma	Σ	σ
eta	H	η	tau	T	τ
theta	Θ	θ	upsilon	Y	υ
iota	I	ι	phi	Φ	φ
kappa	K	κ	chi	X	χ
lambda	Λ	λ	psi	Ψ	ψ
mu	M	μ	omega	Ω	ω

Periodic classification of the elements

Periodic classification appears on pp. 284–5.

Group	I	II

1st Period
                                                                                        1
                                                                                        H

2nd Period   3   4
             Li Be

3rd Period   11 12
             Na Mg

4th Period   19 20
             K   Ca

5th Period   37 38   lanthanides
             Rb Sr

6th Period   55 56   57   58   59   60   61   62   63   64   65   66   67   68   69   70
             Cs Ba   La   Ce   Pr   Nd   Pm   Sm   Eu   Gd   Tb   Dy   Ho   Er   Tm   Yb

7th Period   87 88   89   90   91   92   93   94   95   96   97   98   99   100 101 102
             Fr Ra   Ac   Th   Pa   U   Np   Pu   Am   Cm   Bk   Cf   Es   Fm   Md   No

                                                 actinides

                                                                              typical elements —

										III	IV	V	VI	VII	0
															2 He
										5 B	6 C	7 N	8 O	9 F	10 Ne
										13 Al	14 Si	15 P	16 S	17 Cl	18 Ar
21 Sc	22 Ti	23 V	24 Cr	25 Mn	26 Fe	27 Co	28 Ni	29 Cu	30 Zn	31 Ga	32 Ge	33 As	34 Se	35 Br	36 Kr
39 Y	40 Zr	41 Nb	42 Mo	43 Tc	44 Ru	45 Rh	46 Pd	47 Ag	48 Cd	49 In	50 Sn	51 Sb	52 Te	53 I	54 Xe
71 Lu	72 Hf	73 Ta	74 W	75 Re	76 Os	77 Ir	78 Pt	79 Au	80 Hg	81 Tl	82 Pb	83 Bi	84 Po	85 At	86 Rn
103 Lr	104	105	106												

transition elements

Index